WORKSHEETS

FOR CLASSROOM OR LAB PRACTICE

CHRISTINE VERITY

INTRODUCTORY AND INTERMEDIATE ALGEBRA

FOURTH EDITION

MARVIN BITTINGER

Indiana University Purdue University Indianapolis

JUDITH BEECHER

Indiana University Purdue University Indianapolis

Addison-Wesley
is an imprint of

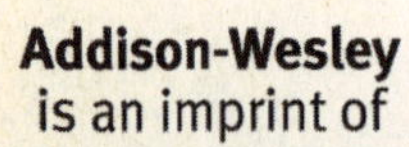

www.pearsonhighered.com

Table of Contents

Chapter 1 INTRODUCTION TO REAL NUMBERS AND ALGEBRAIC EXPRESSIONS

1.1 Introduction to Algebra

Learning Objectives
a   Evaluate algebraic expressions by substitution.
b   Translate phrases to algebraic expressions.

Key Terms
Use the vocabulary terms listed below to complete each statement in Exercises 1–6.

algebraic expression   constant   evaluating

substituting   value   variable

1. A combination of letters, numbers, and operation signs, such as $16x - 18y$ is called a(n)

   _____________________ .

2. A letter that can represent various numbers is a(n) _____________________ .

3. A letter that can stand for just one number is a(n) _____________________ .

4. When we replace a variable with a number, we are _____________________ for the

   variable.

5. When we replace all variables in an expression with numbers and carry out the

   operations, we are _____________________ the expression.

6. The results of evaluating an algebraic expression is called the _____________________ of

   the expression.

Objective a   Evaluate algebraic expressions by substitution.

Substitute to find values of the expressions in each of the following applied problems.

7. The area A of a triangle with base b and height h is given
   by $A = \frac{1}{2}bh$. Find the area when $b = 32$ cm (centimeters)
   and $h = 15$ cm.

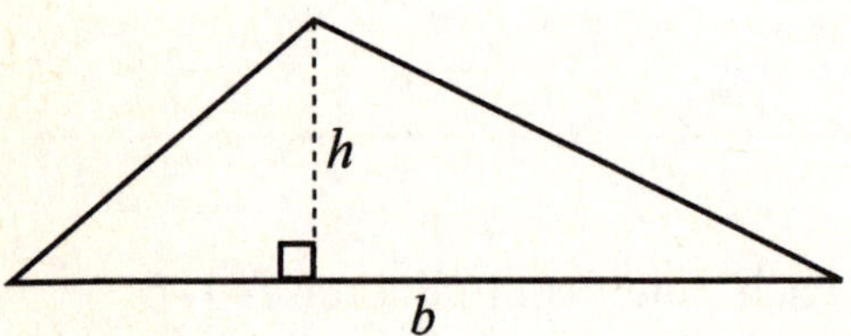

7. _________________

8. A driver who drives at a constant speed of r mph for t hr
   will travel a distance d mi given by $d = rt$ mi. How far
   will a driver travel at the speed of 70 mph for 3 hr?

8. _________________

9. The simple interest on a principal of P dollars at interest
   rate r for time t, in years, is given by $I = Prt$. Find the
   simple interest on a principal of \$1500 at 6% for 3 yr.
   (Hint: 6% = 0.06)

9. _________________

10. A rectangular piece of paper is 5 in. wide and 9 in. long.
    Find its area.

10. _________________

Evaluate.

11. $10z$, when $z = 3$

11. _________________

2

12. $\dfrac{x}{y}$, when $x = 42$ and $y = 6$

12. _________________

13. $\dfrac{5p}{q}$, when $p = 8$ and $q = 10$

13. _________________

14. $\dfrac{a-b}{3}$, when $a = 25$ and $b = 13$

14. _________________

Objective b Translate phrases to algebraic expressions.

Translate each phrase to an algebraic expression. Use any letter for the variable unless directed otherwise.

15. Eight more than some number

15. _________________

16. Twenty less than some number

16. _________________

17. b divided by x

17. _________________

18. c subtracted from a

18. _________________

19. The product of two numbers

19. _________________

20. Six multiplied by some number

20. _________________

21. Ten more than seven times some number

21. _________________

22. Five less than the product of two numbers

23. Four times some number plus seven

24. The sum of twice a number plus five times another number

25. The price of a treadmill after a 25% reduction if the price before the reduction was p

26. Raena drove a speed of 55 mph for t hours. How far did Raena drive? (See Exercise 8.)

22. _______________

23. _______________

24. _______________

25. _______________

26. _______________

4

Chapter 1 INTRODUCTION TO REAL NUMBERS
AND ALGEBRAIC EXPRESSIONS

1.2 The Real Numbers

Learning Objectives

a   State the integer that corresponds to a real-world situation.
b   Graph rational numbers on the number line.
c   Convert from fraction notation for a rational number to decimal notation.
d   Determine which of two real numbers is greater and indicate which, using < or >.
    Given an inequality like $a > b$, write another inequality with the same meaning.
    Determine whether an inequality like $-3 \leq 5$ is true or false.
e   Find the absolute value of a real number.

Key Terms

Use the vocabulary terms listed below to complete each statement in Exercises 1–8.

absolute value   graph   integers   natural numbers

opposites   rational numbers   set   whole numbers

1.   A(n) _________________ is a collection of objects.

2.   We call -1 and 1 _________________ of each other.

3.   To _________________ a number means to find and mark its point on a number

     line.

4.   The _________________ of a number is its distance from zero on a number line.

5.   The set of _________________ = $\{1, 2, 3,...\}$.

6.   The set of _________________ = $\{0, 1, 2, 3,...\}$.

7.   The set of _________________ = $\{..., -4, -3, -2, -1, 0, 1, 2, 3, 4,...\}$.

8.   The set of _________________ = the set of numbers $\frac{a}{b}$, where a and b are integers

     and $b \neq 0$.

5

Objective a   State the integer that corresponds to a real-world situation.

State the integers that correspond to the situation.

9. On March 12, the temperature was 3° below zero. On March 21, it was 55° above zero.

9. _________________

10. Will withdrew $480 from his savings account to buy textbooks. The next day, he deposited his paycheck of $325.

10. _________________

Objective b   Graph rational numbers on a number line.

Graph the number on the number line.

11. $-\dfrac{7}{4}$

11. _________________

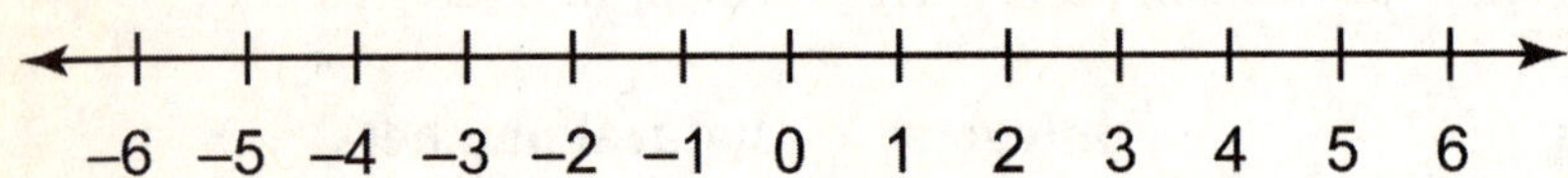

12. 2.67

12. _________________

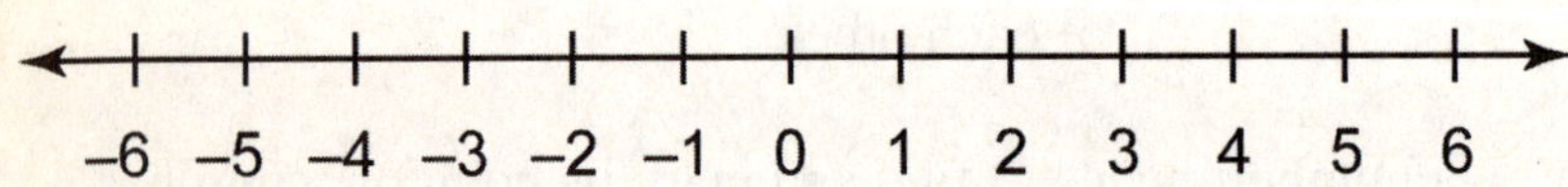

Objective c   Convert from fraction notation for a rational number to decimal notation.

Convert to decimal notation.

13. $-\dfrac{3}{8}$

13. _________________

14. $\dfrac{7}{3}$

14. _________________

6

15. $\dfrac{13}{6}$

15. _______________

16. $-\dfrac{8}{9}$

16. _______________

17. $-\dfrac{5}{4}$

17. _______________

18. $\dfrac{3}{20}$

18. _______________

Objective d   Determine which of two real numbers is greater and indicate which, using < or >. Given an inequality like $a > b$, write another inequality with the same meaning. Determine whether an inequality like $-3 \le 5$ is true or false.

Use either < or > for $\square$ to write a true sentence.

19. $4 \;\square\; 0$

19. _______________

20. $-3 \;\square\; 1$

20. _______________

21. $14 \;\square\; -14$

21. _______________

22. $0 \;\square\; -6$

22. _______________

23. $-8 \;\square\; -9$

23. _______________

24. $-4.6 \;\square\; -1.2$

24. _______________

25. $\dfrac{3}{4} \;\square\; -\dfrac{5}{4}$

25. _______________

26. $\dfrac{2}{5} \,\square\, \dfrac{3}{8}$

Write true or false.

27. $-3 \le -4$

28. $0 \ge -5$

Write an inequality with the same meaning.

29. $-3 \ge x$

30. $5 < y$

Objective e Find the absolute value of a real number.

Find the absolute value.

31. $|-16|$

32. $|12|$

33. $\left|-\dfrac{1}{4}\right|$

34. $|-12.3|$

35. $\left|4\dfrac{7}{8}\right|$

36. $\left|\dfrac{0}{5}\right|$

26. _______________

27. _______________

28. _______________

29. _______________

30. _______________

31. _______________

32. _______________

33. _______________

34. _______________

35. _______________

36. _______________

8

Name: Date:
Instructor: Section:

Chapter 1 INTRODUCTION TO REAL NUMBERS AND ALGEBRAIC EXPRESSIONS

1.3 Addition of Real Numbers

Learning Objectives
a   Add real numbers without using the number line.
b   Find the opposite, or additive inverse, of a real number.
c   Solve applied problems involving addition of real numbers.

Key Terms
Use the vocabulary terms listed below to complete each statement in Exercises 1–4.

additive inverses   negative   positive   zero

1.   The sum of two positive integers is _______________________ .

2.   The sum of two negative integers is _______________________ .

3.   If $a + b = a$, then b must be _______________________ .

4.   Two numbers whose sum is 0 are called _______________________ .

Objective a   Add real numbers without using a number line.

Add. Do not use the number line except as a check.

5. $6 + (-11)$ 5. _______________

6. $-8 + 3$ 6. _______________

7. $-7 + 7$ 7. _______________

8. $27 + (-27)$ 8. _______________

9. $-13+0$

9. _______________

10. $-3+12$

10. _______________

11. $50+(-25)$

11. _______________

12. $-2+(-17)$

12. _______________

13. $-1.3+(-3.8)$

13. _______________

14. $\dfrac{2}{9}+\left(-\dfrac{7}{9}\right)$

14. _______________

15. $-\dfrac{7}{6}+\dfrac{1}{3}$

15. _______________

16. $-\dfrac{3}{20}+\dfrac{7}{15}$

16. _______________

17. $-\dfrac{5}{24}+\left(-\dfrac{7}{18}\right)$

17. _______________

18. $-\dfrac{3}{20}+\left(-\dfrac{7}{24}\right)$

18. _______________

10

19. $-32 + \left(-\dfrac{5}{11}\right) + 108 + \left(-\dfrac{6}{11}\right)$

19. ______________________

20. $36 + (-45) + 657 + (-506) + (-89)$

20. ______________________

Objective b   Find the opposite, or additive inverse, of a real number.

Find the opposite, or additive inverse.

21. -48

21. ______________________

22. 16.3

22. ______________________

Evaluate $-x$ when:

23. $x = \dfrac{11}{16}$

23. ______________________

24. $x = -54$

24. ______________________

Evaluate $-(-x)$ when:

25. $x = -19$

25. ______________________

26. $x = 8.3$

26. ______________________

Find the opposite. (Change the sign.)

27. 28

27. ______________________

28. $-\dfrac{7}{3}$

28. ______________________

Objective c   Solve applied problems involving addition of real numbers.

Solve.

29. In a football game, the home team lost 5 yd on the first
    play and gained 8 yd on the second play. Find the total
    gain or loss.

29. _______________

30. On December 1, LaKenya's credit card bill shows that she
    owes $125. During the month of December, LaKenya
    sends a check for $95 to the credit card company, charges
    another $860 for gifts, and then pays off another $400 of
    her bill. What is the new balance of LaKenya's account at
    the end of December?

30. _______________

12

Chapter 1 INTRODUCTION TO REAL NUMBERS AND ALGEBRAIC EXPRESSIONS

1.4 Subtraction of Real Numbers

Learning Objectives
a   Subtract real numbers and simplify combinations of additions and subtractions.
b   Solve applied problems involving subtraction of real numbers.

Key Terms
Use the vocabulary terms listed below to complete each statement in Exercises 1–2.

difference   **opposite**

1. The _________________ $a-b$ is the number c for which $a=b+c$.

2. To subtract, add the _________________ of the number being subtracted.

Objective a   Subtract real numbers and simplify combinations of additions and subtractions.

Subtract.

3. $4-14$

3. _________________

4. $-10-(-10)$

4. _________________

5. $-2-(-17)$

5. _________________

6. $13-20$

6. _________________

7. $3-(-3)$

7. _________________

8. $18-(-40)$

8. _________________

9. $0-(-7)$

9. _________________

10. $-16-0$

10. _________________

13

11. $-\dfrac{1}{6}-\dfrac{11}{12}$

11. _______________

12. $-\dfrac{1}{9}-\left(-\dfrac{5}{6}\right)$

12. _______________

13. $2.6-(-5.2)$

13. _______________

14. $8-4.63$

14. _______________

15. $3-(-4.75)$

15. _______________

16. $0-(-200)$

16. _______________

Simplify.

17. $32-(-25)-36+17-9$

17. _______________

18. $64-98+(-17)-(-38)-95$

18. _______________

19. $-1.7-(-15.7)+(-12.3)-(-16)$

19. _______________

20. $-\dfrac{2}{9}+\dfrac{5}{6}-\left(-\dfrac{7}{12}\right)$

20. _______________

Objective b   Solve applied problems involving subtraction of real numbers.

Solve.

21. From an elevation of 38 ft below sea level, Devin climbed to an elevation of 92 ft above sea level. Find the difference in elevations.

21. _______________

22. On January 10, the temperature fell from 12°F to –8°F. Find the difference between these two temperatures.

22. _______________

14

Chapter 1 INTRODUCTION TO REAL NUMBERS AND ALGEBRAIC EXPRESSIONS

1.5 Multiplication of Real Numbers

Learning Objectives
a   Multiply real numbers.
b   Solve applied problems involving multiplication of real numbers.

Key Terms
Use the vocabulary terms listed below to complete each statement in Exercises 1-4. Each term will be used more than once.

negative   positive

1.   The product of a positive number and a negative number is _____________________ .

2.   The product of two negative numbers is _____________________ .

3.   The product of an even number of negative numbers is _____________________ .

4.   The product of an odd number of negative numbers is _____________________ .

Objective a   Multiply real numbers.

Multiply.

5. $-5 \cdot 4$

5. _____________________

6. $6 \cdot (-5)$

6. _____________________

7. $-8 \cdot (-7)$

7. _____________________

8. -11×12

8. _____________________

9. $-6.7 \cdot (-32)$

9. _____________________

10. $-84 \cdot (-4.5)$

10. _____________________

15

11. $-\dfrac{5}{9}\cdot\left(\dfrac{3}{10}\right)$

12. $-\dfrac{4}{5}\cdot\left(\dfrac{7}{8}\right)$

13. $-2\cdot(-5)\cdot(-8)\cdot(-1)$

14. $0.02\cdot(-4)\cdot8\cdot(-5)$

15. $5\cdot(-4)\cdot3\cdot(-2)\cdot1\cdot0$

16. $(-2)(-3)(-4)(-5)(-6)$

17. Evaluate $(-5x)^2$ and $-5x^2$ when $x=-4$.

18. Evaluate $3x^2$ when $x=6$ and when $x=-6$.

Objective b   Solve applied problems involving multiplication of real numbers.

Solve.

19. Vivian lost 1.5 lb each week for a period of 8 weeks. Express her total weight change as an integer.

20. The temperature of a chemical compound was $-10°C$ at 4:30 p.m. During a reaction, it dropped $2°C$ per minute until 4:38 P.M. What was the temperature at 4:38 P.M.?

21. The price of a stock began the day at $42.75 per share and dropped $0.38 per hour for 8 hours. What was the price of the stock after 8 hr?

22. After diving 55 m below sea level, a diver rises at a rate of 6 m per minute for 8 min. Where is the diver in relation to the surface at the end of the 8-min period?

11. ______________

12. ______________

13. ______________

14. ______________

15. ______________

16. ______________

17. ______________

18. ______________

19. ______________

20. ______________

21. ______________

22. ______________

16

Chapter 1 INTRODUCTION TO REAL NUMBERS
AND ALGEBRAIC EXPRESSIONS

1.6 Division of Real Numbers

Learning Objectives
a   Divide integers.
b   Find the reciprocal of a real number.
c   Divide real numbers.
d   Solve applied problems involving division of real numbers.

Key Terms
Use the vocabulary terms listed below to complete each statement in Exercises 1-4.

negative   **not defined**   **positive**   **reciprocals**

1. Division by 0 is _____________________ .

2. The quotient of two numbers with the same sign is _____________________ .

3. The quotient of two numbers with different signs is _____________________ .

4. Two numbers whose product is 1 are called _____________________ .

Objective a   Divide integers.

Divide, if possible. Check each answer.

5. $36 \div (-4)$ 5. _______________

6. $\dfrac{26}{-2}$ 6. _______________

7. $\dfrac{-75}{-3}$ 7. _______________

8. $-66 \div (-6)$

8. ________________

9. $\dfrac{66}{0}$

9. ________________

10. $\dfrac{0}{-3}$

10. ________________

Objective b   Find the reciprocal of a real number.

Find the reciprocal.

11. $\dfrac{4}{9}$

11. ________________

12. $-\dfrac{6}{17}$

12. ________________

13. -12

13. ________________

14. 4.6

14. ________________

15. $\dfrac{-1}{3t}$

15. ________________

16. $\dfrac{-2x}{3y}$

16. ________________

Objective c   Divide real numbers.

Rewrite the division as a multiplication.

17. $3 \div 8$

17. ________________

18. $\dfrac{5}{-9}$

18. ________________

18

19. $-\dfrac{6.8}{1.3}$

20. $\dfrac{c}{\dfrac{1}{d}}$

21. $\dfrac{2x-1}{4}$

22. $\dfrac{3a+a^2}{a+1}$

Divide.

23. $\dfrac{5}{6}\div\left(-\dfrac{7}{5}\right)$

24. $-\dfrac{3}{8}\div\left(-\dfrac{6}{7}\right)$

25. $-\dfrac{2}{5}\div\left(-\dfrac{4}{9}\right)$

26. $-\dfrac{2}{3}\div\left(-\dfrac{3}{2}\right)$

27. $-10.4\div 2.6$

19. _______________

20. _______________

21. _______________

22. _______________

23. _______________

24. _______________

25. _______________

26. _______________

27. _______________

28. $\dfrac{-2}{11-11}$

28. _______________

Objective d   Solve applied problems involving division of real numbers.

A percent of increase is generally positive and a percent of decrease is generally negative. In Exercises 29 and 30, find the missing numbers. Round to the nearest tenth of a percent.
(Source: Veronis Shuler Stevenson, New York, New York)

29.

29. _______________

Hours Per Person Spent Playing Video Games			
2000	2008	Change	Percent of Increase or Decrease
59	98	39	

30.

30. _______________

Hours Per Person Spent Reading Magazines			
2000	2008	Change	Percent of Increase or Decrease
135	110	−25	

20

Chapter 1 INTRODUCTION TO REAL NUMBERS AND ALGEBRAIC EXPRESSIONS

1.7 Properties of Real Numbers

Learning Objectives
a   Find equivalent fraction expressions and simplify fraction expressions.
b   Use the commutative and associative laws to find equivalent expressions.
c   Use the distributive laws to multiply expressions like 8 and $x - y$.
d   Use the distributive laws to factor expressions like $4x - 12 + 24y$.
e   Collect like terms.

Key Terms
Use the vocabulary terms listed below to complete each statement in Exercises 1–7.

associative   commutative   distributive

equivalent   factor   like   terms

1.   Two expressions that have the same value for all allowable replacements are called

     _____________________ expressions.

2.   The statement $3 + x = x + 3$ illustrates a(n) _____________________ law.

3.   The statement $(2p)q = 2(pq)$ illustrates a(n) _____________________ law.

4.   The statement $5(a + b) = 5a + 5b$ illustrates a(n) _____________________ law.

5.   The _____________________ of an expression are separated by addition signs.

6.   To _____________________ an expression, we find an equivalent expression that is a

     product.

7.   _____________________ terms have exactly the same variable factors.

Objective a   Find equivalent fraction expressions and simplify fraction expressions.

Find an equivalent expression with the given denominator.

8. $\dfrac{2}{7} = \dfrac{\square}{14x}$

9. $\dfrac{3}{10y} = \dfrac{\square}{10xy}$

Simplify.

10. $-\dfrac{35xy}{25xy}$

11. $\dfrac{48ab}{40a}$

Objective b   Use the commutative and associative laws to find equivalent expressions.

Write an equivalent expression. Use a commutative law.

12. $t + 5$

13. $18 + xy$

Write an equivalent expression. Use an associative law.

14. $3 + (x + y)$

15. $(7a)b$

8. _______________

9. _______________

10. _______________

11. _______________

12. _______________

13. _______________

14. _______________

15. _______________

22

Use the commutative and associative laws to write three equivalent expressions.

16. $4+(m+n)$

16. _______________

17. $(xy)9$

17. _______________

Objective c   **Use the distributive laws to multiply expressions like 8 and $x - y$.**

Multiply.

18. $3(2x+8)$

18. _______________

19. $5(x+7+2n)$

19. _______________

20. $11(p-3)$

20. _______________

21. $\dfrac{3}{4}(y-12)$

21. _______________

22. $-4(-6x-y+9)$

22. _______________

23. $-1.4(-2.3a-1.6b-3.4)$

23. _______________

List the terms of each expression.

24. $2a-1.4b+10c$

24. _______________

25. $5x+6y-7z$

25. _______________

Objective d   **Use the distributive laws to factor expressions like $4x-12+24y$.**

Factor. Check by multiplying.

26. $3x+18$

26. _______________

27. $12y+30z$

27. _______________

23

28. $9m - 45$

28. _______________

29. $-5t + 20$

29. _______________

30. $-21c + 35d + 7$

30. _______________

31. $\dfrac{3}{4}x - \dfrac{7}{4}y + \dfrac{1}{4}$

31. _______________

Objective e   Collect like terms.

Collect like terms.

32. $14x - x$

32. _______________

33. $8t - 11t$

33. _______________

34. $4m^2 - 10m - 5m^2$

34. _______________

35. $16 - 40c - 20 - 5c + 6 + 8c$

35. _______________

36. $-4 + 8a - 5b + 36a - 17b + 12$

36. _______________

37. $5y + z + \dfrac{1}{3}y - \dfrac{3}{5}z - 11$

37. _______________

24

Chapter 1 INTRODUCTION TO REAL NUMBERS
AND ALGEBRAIC EXPRESSIONS

1.8 Simplifying Expressions; Order of Operations

Learning Objectives
a   Find an equivalent expression for an opposite without parentheses, where an expression has several terms.
b   Simplify expressions by removing parentheses and collecting like terms.
c   Simplify expressions with parentheses inside parentheses.
d   Simplify expressions using rules for order of operations.

Key Terms
Exercises 1–4 list the rules for order of operations. Use the vocabulary terms listed below to complete each statement.

   additions   divisions   exponential   grouping

1.   Do all calculations within _____________________ symbols before operations outside.

2.   Evaluate all _____________________ expressions.

3.   Do all multiplications and _____________________ in order from left to right.

4.   Do all _____________________ and subtractions in order from left to right.

Objective a   Find an equivalent expression for an opposite without parentheses, where an expression has several terms.

Find an equivalent expression without parentheses.

5. $-(3x + y - 10z)$ 5. _______________

6. $-(a - 4b + 6c)$ 6. _______________

7. $-(-10m - 3n - 18)$ 7. _______________

8. $-(-2p + 5q - 8t)$ 8. _______________

Objective b   Simplify expressions by removing parentheses and collecting like terms.

Remove parentheses and simplify.

9. $6x - (2x + 7)$

9. ______________

10. $5y + 2y - (3y + 1)$

10. ______________

11. $4p - 5q - 2(3p - 9q)$

11. ______________

12. $12m - n - 3(5m - 2n + 6p)$

12. ______________

13. $(5a + 9b) - 3(4a - 2b)$

13. ______________

14. $(15x + 7y - 8z) - 4(-4x - 2y + 10z)$

14. ______________

Objective c   Simplify expressions with parentheses inside parentheses.

Simplify.

15. $6[3 - 2(7 - 4)]$

15. ______________

16. $[5(11 - 2) + 7] - [9 - (3 + 5)]$

16. ______________

17. $[4(x + 5) - 10] + [3(x - 6) + 7]$

17. ______________

26

18. $\left[8(x+2)-12\right]-\left[10(x-3)+1\right]$

18. ___________________

19. $5\left\{\left[3(x-4)+8\right]-\left[6(2x-1)+9\right]\right\}$

19. ___________________

20. $2\left\{\left[3(x-6)+5\right]-4\left[5(x+2)-7\right]\right\}$

20. ___________________

Objective d   Simplify expressions using rules for order of operations.

Simplify.

21. $4-3\cdot5-6$

21. ___________________

22. $\left[(-28)\div(-7)\right]\div\left(-\dfrac{1}{2}\right)$

22. ___________________

23. $13-3^2+3^3$

23. ___________________

24. $2^4+13\cdot16-(12+32\cdot3)$

24. ___________________

25. $6\cdot7-2\cdot4+10$

25. ___________________

26. $10^2/20$

26. ___________________

27. $35-4(-5)+8$

27. ___________________

28. $15 \div (-3) + 20 \div 2$

28. ______________

29. $-7^2 - 8$

29. ______________

30. $26 - 30^2$

30. ______________

31. $3 \cdot 5^2 - 65$

31. ______________

32. $\dfrac{4 - 3^2}{2^3 + 2^2}$

32. ______________

33. $\dfrac{2(8-9) - 3 \cdot 4}{5 \cdot 6 - 3(8-1)}$

33. ______________

34. $\dfrac{\left|2^3 - 4^2\right| + \left|10 \cdot 11\right|}{-24 \div (-3) \div (-4)}$

34. ______________

28

Chapter 2 SOLVING EQUATIONS AND INEQUALITIES

2.1 Solving Equations: The Addition Principle

Learning Objectives
a   Determine whether a given number is a solution of a given equation.
b   Solve equations using the addition principle.

Key Terms
Use the vocabulary terms listed below to complete each statement in Exercises 1–4.

   addition principle   **equation**   **equivalent equations**   **solution**

1.   A(n) ___________________ is a number sentence that says that the expressions on

   either side of the equals sign represent the same number.

2.   Any replacement for the variable that makes an equation true is called

   a(n) ___________________ of the equation.

3.   Equations with the same solutions are called ___________________ .

4.   The ___________________ states that for any real numbers a, b, and c, $a = b$ is

   equivalent to $a + c = b + c$.

Objective a   Determine whether a given number is a solution of a given equation.

Determine whether the given number is a solution of the given equation.

5.   13; $x + 19 = 22$ 5. _____________

6.   23; $t - 19 = 4$ 6. _____________

7. $-6;\ 9t=-54$

7. _______________

8. $56;\ \dfrac{y}{8}=9$

8. _______________

9. $14;\ 5x+3=73$

9. _______________

10. $-9;\ 3(x-1)=24$

10. _______________

Objective b   Solve equations using the addition principle.

Solve using the addition principle. Don't forget to check!

11. $x+8=20$

11. _______________

Check: $\underline{x+8=20}$

$\ \ \ \ \ \ \ \ \ |$

$\ \ \ \ \ \ \ \ ?$

$\ \ \ \ \ \ \ \ \ |$

12. $t+12=42$

12. _______________

Check: $\underline{t+12=42}$

$\ \ \ \ \ \ \ \ \ |$

$\ \ \ \ \ \ \ \ ?$

$\ \ \ \ \ \ \ \ \ |$

30

13. $z + 3 = -8$

13. ______________

14. $y - 5 = 21$

14. ______________

15. $a - 3 = -5$

15. ______________

16. $-6 + x = 17$

16. ______________

17. $-8 + y = -15$

17. ______________

18. $a + \dfrac{2}{3} = 12$

18. ______________

19. $y - \dfrac{3}{5} = \dfrac{7}{8}$

19. ______________

20. $10.8 = x + 4.9$

20. ______________

21. $-2.9 = -1.4 + y$

21. ______________

22. $6\dfrac{2}{3} = 4\dfrac{1}{2} + x$

22. ______________

Chapter 2 SOLVING EQUATIONS AND INEQUALITIES

2.2 Solving Equations: The Multiplication Principle

Learning Objective
a   Solve equations using the multiplication principle.

Key Terms
Use the vocabulary terms listed below to complete each statement in Exercises 1–4.

coefficient   **identity**   **inverse**   **principle**

1. The multiplication ____________________ states that for any real numbers a, b, and c, $c \neq 0$, $a = b$ is equivalent to $a \cdot c = b \cdot c$.

2. The multiplicative ____________________ of 3 is $\dfrac{1}{3}$.

3. The multiplicative ____________________ is 1 since $1 \cdot x = x$.

4. The ____________________ of $8x$ is 8.

Objective a   Solve equations using the multiplication principle.

Solve using the multiplication principle. Don't forget to check!

5. $4x = 28$ 5. ________________

6. $3x = 36$ 6. ________________

7. $-y = 20$ 7. ________________

8. $-t = -5$ 8. ________________

9. $11x = -66$ 9. ________________

33

10. $-9x = -108$

11. $-32x = -96$

12. $\dfrac{m}{8} = -7$

13. $\dfrac{2}{5}y = 18$

14. $\dfrac{-a}{7} = 8$

15. $-\dfrac{t}{4} = \dfrac{1}{3}$

16. $-\dfrac{3}{4}r = \dfrac{5}{8}$

17. $-\dfrac{4}{5}a = -\dfrac{16}{15}$

18. $5.9x = 17.7$

19. $-3.6y = -21.6$

20. $-\dfrac{3}{8}m = -22.14$

21. $\dfrac{-x}{3} = -45$

22. $\dfrac{t}{-6} = 8$

10. _______________

11. _______________

12. _______________

13. _______________

14. _______________

15. _______________

16. _______________

17. _______________

18. _______________

19. _______________

20. _______________

21. _______________

22. _______________

34

Chapter 2 SOLVING EQUATIONS AND INEQUALITIES

2.3 Using the Principles Together

> **Learning Objectives**
> a   Solve equations using both the addition and the multiplication principles.
> b   Solve equations in which like terms may need to be collected.
> c   Solve equations by first removing parentheses and collecting like terms; solve equations
>   with an infinite number of solutions and equations with no solutions.

Key Terms
Use the vocabulary terms listed below to complete each statement in Exercises 1–4.

clear fractions   **distributive laws**   **infinitely many solutions**

no solution

1.   We remove parentheses in an equation by multiplying using the ___________________.

2.   We multiply every term on both sides of an equation by the least common multiple of all

     denominators in order to ___________________ .

3.   When solving an equation, if we end with a true equation, the equation has

     ___________________ .

4.   When solving an equation, if we end with a false equation, the equation has

     ___________________ .

Objective a   Solve equations using both the addition and the multiplication principles.

Solve. Don't forget to check!

5. $3x + 5 = 29$ 5. ___________________

6. $6x - 5 = 37$ 6. ___________________

7. $4x + 5 = -39$

7. _______________

8. $-23 = 7 + 5y$

8. _______________

9. $-6x + 11 = 29$

9. _______________

10. $-7x - 18 = -28\dfrac{1}{2}$

10. _______________

Objective b   Solve equations in which like terms may need to be collected.

Solve.

11. $4x + 5x = 63$

11. _______________

12. $6x + 5x = 132$

12. _______________

13. $-3y - 4y = 28$

13. _______________

14. $x + \dfrac{1}{2}x = 12$

14. _______________

36

15. $7x - 1 = 15 - x$ 15. _______________

16. $8x + 7 = 3x + 12$ 16. _______________

17. $2 + 5z - 11 = 5z + 4 - z$ 17. _______________

18. $4y - 3 + 2y = 6y + 8 - y$ 18. _______________

Solve. Clear fractions or decimals first.

19. $\dfrac{3}{4}x + \dfrac{1}{2}x = 5x + \dfrac{1}{2} + \dfrac{1}{4}x$ 19. _______________

20. $\dfrac{1}{6} + 2y = 7y - \dfrac{5}{12}$ 20. _______________

21. $\dfrac{2}{5} + \dfrac{3}{5}x = \dfrac{8}{15} + \dfrac{1}{2}x + \dfrac{3}{2}$ 21. _______________

22. $3.6x + 14.7 = 0.3 - 1.2x$ 22. _______________

23. $4.07 - 0.61x = 0.82 - 5.16x$ 23. _______________

24. $\dfrac{2}{3}x - \dfrac{1}{4}x = \dfrac{3}{5}x + 1$ 24. _______________

Objective c Solve equations by first removing parentheses and collecting like terms; solve equations with an infinite number of solutions and equations with no solutions.

Solve.

25. $4(2t-5)=28$

25. _______________

26. $-13+x=x+15$

26. _______________

27. $3(2+5y)-10=11$

27. _______________

28. $8-5(2x-3)=3$

28. _______________

29. $4x+3-5x-18=6-8x+7x-21$

29. _______________

30. $3(m+2)=8(m+7)$

30. _______________

31. $10(3t+2)=7(4t+6)$

31. _______________

32. $13-(3x+4)=5(x+7)+x$

32. _______________

33. $13x-2-7x=2(3x-1)+5$

33. _______________

34. $2\left[5-3(4-x)\right]-7=3\left[2(5x-1)+8\right]-15$

34. _______________

38

Chapter 2 SOLVING EQUATIONS AND INEQUALITIES

2.4 Formulas

Learning Objectives
a   Evaluate a formula.
b   Solve a formula for a specified letter.

Key Terms
Use the vocabulary terms listed below to complete each statement in Exercises 1–2.

evaluating   **formula**

1.   A(n) _____________________ is an equation relating several quantities.

2.   When we replace the variables in an expression with numbers and calculate the result,

     we are _____________________ the expression.

Objective a   Evaluate a formula.

Solve.

3.   The formula $d = 65t$ gives the distance traveled, in miles,   **3. a)**________________
     by a vehicle traveling 65 mph for t hr.
     a)   A car travels 65 mph for 3 hr. How many miles did   **b)**________________
          the car travel?

     b)   Solve for t.

4. The formula $A = lw$ gives the area of a rectangle with length l and width w.

   a) A rectangle has a length of 4 m and a width of $\frac{1}{2}$ m. What is the area of the rectangle?

   b) Solve for w.

4. a)_________________

   b)_________________

5. The formula $A = bh$ gives the area of a parallelogram with base b and height h. A parallelogram has base 14 cm and height 9 cm. What is the area of the parallelogram?

5. _________________

6. The cost for one month of Camden's cell phone, in dollars, is given by the formula $c = 45 + 0.1m$, where m is the number of text messages sent or received that month. How much was his cell phone bill for a month in which he sent or received 80 text messages?

6. _________________

40

Objective b   Solve a formula for a specified letter.

Solve for the indicated letter.

7. $c = 8d$, for d 7._______________

8. $n = 8 + m$, for m 8._______________

9. $y = 18 - x$, for x 9._______________

10. $11a = 12b$, for b 10._______________

11. $P = ax + b$, for x 11._______________

12. $A = \dfrac{p + q + r}{3}$, for q 12._______________

13. $A = 4\pi r^2$, for r^2

13. _______________

14. $M = \dfrac{x - y}{2}$, for x

14. _______________

15. $c = \dfrac{4k}{w}$, for w

15. _______________

16. $t = \dfrac{xy}{z}$ for z

16. _______________

42

Chapter 2 SOLVING EQUATIONS AND INEQUALITIES

2.5 Applications of Percent

Learning Objective
a Solve applied problems involving percent.

Key Terms
Use the vocabulary terms listed below to complete each statement in Exercises 1–4.

%	is	of	what number

1. _____________________ translates to "·" or "×".

2. _____________________ translates to "=".

3. _____________________ translates to any letter.

4. _____________________ translates to "$\times \dfrac{1}{100}$" or "$\times 0.01$".

Objective a   Solve applied problems involving percent.

Solve.

5. What percent of 240 is 156?   5. _______________

6. What percent of 60 is 9?   6. _______________

7. 4.5 is 25% of what number?   7. _______________

8. 72 is 45% of what number?

8. _______________

9. What number is 55% of 280?

9. _______________

10. What number is 24% of 50?

10. _______________

11. What percent of 450 is 72?

11. _______________

12. 0.9 is 15% of what number?

12. _______________

13. 7 is 2% of what number?

13. _______________

14. What percent of 80 is 100?

14. _______________

15. 35 is 140% of what number?

15. _______________

16. What percent of 85 is 85?

16. _______________

44

17. In 2005, shoppers spent $9 billion on Father's Day gifts and $14 billion on Mother's Day gifts. What percent of the amount spent for Mother's Day was spent for Father's Day? Round to the nearest tenth of a percent.
Source: National Retail Federation

17. ______________________

18. In 2006, David Ortiz had 160 hits. His batting average was 0.287, or 28.7%. That is, of the total number of at-bats, 28.7% were hits. How many at-bats did he have?

18. ______________________

19. Chad left a $6 tip for a meal that cost $30.
   a) What percent of the cost of the meal was the tip?

19. a)______________________

b)______________________

b) What was the total cost of the meal including the tip?

20. Claire left a 15% tip of $3.66 for a meal.
   a) What was the cost of the meal before the tip?

20. a)______________________

b)______________________

b) What was the total cost of the meal including the tip?

45

21. The maximum healthy body fat percentage for a woman is 26%. If a woman weighs 150 lb, what is the maximum number of pounds of fat she should have?

21. ________________

22. From 1995 to 2005, U.S. bicycle sales increased by 25%. If sales in 1995 were 15 million, what were the sales in 2005?
Source: National Bicycle Dealers Association

22. ________________

23. The number of people who walk to work decreased from 4.5 million in 1990 to 3.7 million in 2000. What was the percent of decrease?
Source: U.S. Census Bureau; Bureau of Transportation Statistics

23. ________________

24. The number of people who ride a bicycle to work increased from 467,000 in 1990 to 488,000 in 2000. What was the percent of increase?
Source: U.S. Census Bureau; Bureau of Transportation Statistics

24. ________________

46

Chapter 2 SOLVING EQUATIONS AND INEQUALITIES

2.6 Applications and Problem Solving

Learning Objective
a   Solve applied problems by translating to equations.

Key Terms
Use the vocabulary terms listed below to complete each statement in Exercises 1–5.

    check   **familiarize**   **solve**   **state**   **translate**

1.   To ___________________ yourself with a problem, read it carefully, choose a variable

      to represent the unknown, and make a drawing.

2.   To ___________________ a problem into mathematical language, write an equation.

3.   To ___________________ an equation, find all replacements that make the equation

      true.

4.   Always ___________________ the answer in the original problem.

5.   As a final problem-solving step, ___________________ the answer to the problem

      clearly.

Objective a   Solve applied problems by translating to equations.

Solve.

6.   A 60-in. board is cut into two pieces. One piece is four   6. _________________
      times the length of the other. Find the lengths of the
      pieces.

7. Caedan purchased five copies of his favorite CD to give to his friends. If the total spent was $72.80, how much was one copy of the CD?

7. ______________

8. In a recent year, New Hampshire had 117 women holding legislative office. This was 53 more than the number of women holding office in Maryland. How many women held legislative office in Maryland?
Source: U.S. Census Bureau

8. ______________

9. A total of 2 in. of precipitation was recorded in East Lake City on May 11 and 12. The amount recorded on May 11 was three times the amount recorded on May 12. How much was recorded on May 12?

9. ______________

10. The sum of three consecutive integers is 69. What are the numbers?

10. ______________

11. The sum of three consecutive even integers is 198. What are the integers?

11. ______________

48

12. A rectangle has a perimeter of 88 ft. The length is 2 ft
more than twice the width. Find the dimensions of the
rectangle.

12. _______________

13. Caitlyn paid $54.40 for a sweater during a 15%-off sale.
What was the regular price?

13. _______________

14. Carlee paid $27.03, including 6% tax, for decorations for a
party. What was the cost of the decorations before tax?

14. _______________

15. The second angle of a triangle is twice as large as the first
angle. The third angle is 12° more than four times the first
angle. How large are the angles?

15. _______________

16. The balance in Clayton's charge card account grew 3%, to
$669.50, in one month. What was his balance at the
beginning of the month?

16. _______________

17. To mail a first-class package in 2007 cost $0.41 for the first ounce and $0.17 for each additional ounce. If Chelsea paid $1.77 to mail a first-class package, how much did it weigh?

17. _______________

18. A taxi cost $1.50 plus 60¢ per mile. How far can Carissa travel for $8.70?

18. _______________

19. Craig left an 18% tip for a meal. The total cost of the meal, including the tip, was $51.92. What was the cost of the meal before the tip was added?

19. _______________

20. Cayla paid an average of $15 per book for three books. The price of one book was $1 more than another, and the remaining book cost $12. What were the prices of the other two books?

20. _______________

Chapter 2 SOLVING EQUATIONS AND INEQUALITIES

2.7 Solving Inequalities

> **Learning Objectives**
> a   Determine whether a given number is a solution of an inequality.
> b   Graph an inequality on a number line.
> c   Solve inequalities using the addition principle.
> d   Solve inequalities using the multiplication principle.
> e   Solve inequalities using the addition and multiplication principles together.

Key Terms

Use the vocabulary terms listed below to complete each statement in Exercises 1–4.

equivalent   graph   inequality   set-builder notation

1. A(n) ___________________ is a number sentence with $<$, $>$, $\leq$, or $\geq$ as its verb.

2. A(n) ___________________ of an inequality is a drawing that represents its solutions.

3. The sentences $x + 4 < 10$ and $x < 6$ are ___________________ since they have the same solution set.

4. The solution set $\{x \mid x > 2\}$ is written using ___________________ .

Objective a   Determine whether a given number is a solution of an inequality.

Determine whether each number is a solution of the given inequality.

5. $x \leq -6$ 5.

   a) 0   **a)** _______________

   b) −3   **b)** _______________

   c) −6   **c)** _______________

   d) −9   **d)** _______________

   e) −5.4   **e)** _______________

6. $x > 10$

   a) 0

   b) −12

   c) 18

   d) 12.7

   e) 10

6.

   a) _______________

   b) _______________

   c) _______________

   d) _______________

   e) _______________

Objective b   Graph an inequality on a number line.

Graph on a number line.

7. $m < 2$

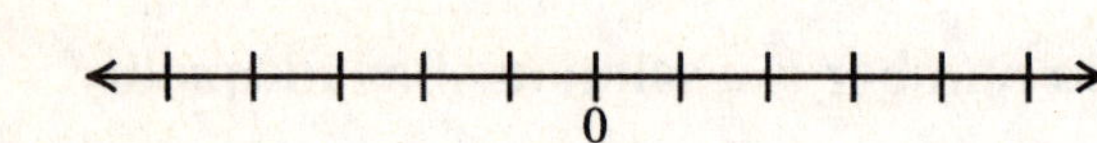

7. _______________

8. $t \geq -2$

8. _______________

9. $-2 \leq x < 5$

9. _______________

10. $-4 < x < 0$

10. _______________

52

Objective c Solve inequalities using the addition principle.

Solve using the addition principle. Then graph.

11. $x + 4 < 3$ 11._____________

12. $x + 3 \geq -2$ 12._____________

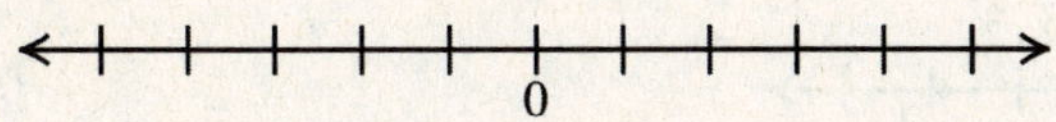

Solve using the addition principle.

13. $t - 8 > 17$ 13._____________

14. $2x + 7 \leq 3x + 2$ 14._____________

15. $-7 + y \geq 14$ 15._____________

16. $x - \dfrac{1}{5} < \dfrac{1}{10}$ 16._____________

Solve using the multiplication principle. Then graph.

17. $-3x \geq 6$

17. _________________

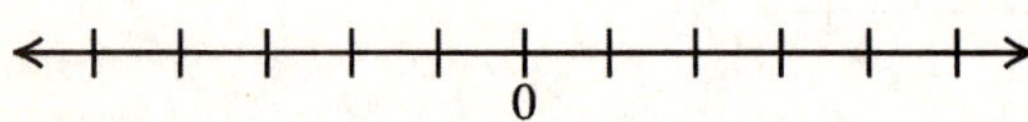

18. $-11x < -44$

18. _________________

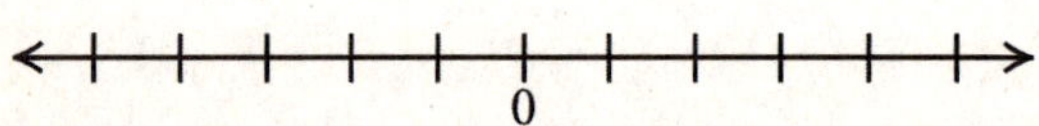

Solve using the multiplication principle.

19. $3x \leq -18$

19. _________________

20. $-5t > -13$

20. _________________

21. $-4p < 7$

21. _________________

22. _________________

22. $-\dfrac{2}{3} \geq 12x$

Name: Date:
Instructor: Section:

Objective e   Solve inequalities using the addition and multiplication principles together.

Solve using the addition and multiplication principles.

23. $2 + 3y > 17$ **23.** ______________

24. $11x - 18 < -51$ **24.** ______________

25. $7x + 8 - 6x \geq 12$ **25.** ______________

26. $12 - 10y \leq 3 - 9y$ **26.** ______________

27. $13 - 8y - y < 40$ **27.** ______________

28. $\dfrac{t}{4} - 6 > 2$

28. ___________

29. $4(3y - 5) \le 4$

29. ___________

30. $2(r - 4) + 5 \ge 7(r + 8) - 16$

30. ___________

56

Chapter 2 SOLVING EQUATIONS AND INEQUALITIES

2.8 Applications and Problem Solving with Inequalities

> **Learning Objectives**
> a   Translate number sentences to inequalities.
> b   Solve applied problems using inequalities.

Key Terms
In Exercises 1–6, match each phrase with its translation from the column on the right. Letters may be used more than once or not at all.

1. _____ x is at least 5

2. _____ x is at most 5

3. _____ x cannot exceed 5

4. _____ x must exceed 5

5. _____ x is no more than 5

6. _____ x is no less than 5

a) $x < 5$
b) $x \leq 5$
c) $x > 5$
d) $x \geq 5$

Objective a   Translate number sentences to inequalities.

Translate to an inequality.

7. A number is at most 13.   7. ________________

8. The coffee weighs more than 2 lb.   8. ________________

9. The speed of the car was between 45 and 65 mph. 9. ________________

10. A number is greater than 10.   10. ________________

11. A number is less than or equal to 50.

11. _________________

12. The cost is at least $45.60.

12. _________________

13. The amount spent on advertising is not to exceed $1500.

13. _________________

14. Five more than one half of a number is no less than −16.

14. _________________

Objective b Solve applied problems using inequalities.

Solve.

15. A student is taking a history course in which five tests are given. To get a C, he must average at least 70 on the five tests. The student got scores of 74, 82, 60, and 68 on the first four tests. Determine (in terms of an inequality) what scores on the last test will allow him to get at least a C.

15. _________________

16. Tin stays solid at Fahrenheit temperatures below $449.7°$. Use the formula $F = \frac{9}{5}C + 32$ to determine (in terms of an inequality) those Celsius temperatures for which tin stays solid.

16. _________________

58

17. To print business cards, Timeless Printing charges a \$25 design fee plus \$0.20 per card. Damon can spend no more than \$85 for business cards. What amount of cards will allow him to stay within budget?

17. _______________

18. Danae has saved \$1500 for college tuition. If her local community college charges a \$65 registration fee plus \$395 per course, what is the greatest number of courses for which Danae can register?

18. _______________

19. The width of a rectangle is fixed at 4 ft. What lengths will make the perimeter at least 20 ft? at most 20 ft?

19. _______________

20. The width of a rectangle is fixed at 8 cm. For what lengths will the area be less than 96 cm²?

20. _______________

21. On August 1, Tom's pond was 18 ft deep. Since that date,   **21.** 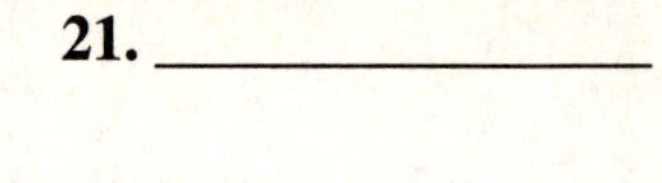

the water level has dropped $\frac{3}{4}$ foot per week. For what

dates will the water level not exceed 15 feet?

22. As part of the requirements for a class, Hannah must   **22.** ________________
volunteer at a community shelter for 4 weeks with an
average of at least 12 hr per week. For the first 3 weeks,
Hannah volunteered 15 hr, 6 hr, and 11 hr. How many hr
must she volunteer in the fourth week in order to meet the
requirement?

60

Chapter 3 GRAPHS OF LINEAR EQUATIONS

3.1 Introduction to Graphing

> **Learning Objectives**
> a   Plot points associated with ordered pairs of numbers; determine the quadrant in which a point lies.
> b   Find the coordinates of a point on a graph.
> c   Determine whether an ordered pair is a solution of an equation with two variables.

Key Terms
Use the vocabulary terms listed below to complete each statement in Exercises 1–5.

    axes   **coordinates**   **graph**   **ordered pair**   **origin**

1.   We graph number pairs on a plane using two perpendicular number lines called

      _____________________.

2.   On the plane, the perpendicular number lines cross at a point called the

      _____________________.

3.   The numbers in an ordered pair are called _____________________.

4.   The notation $(3, -2)$ is an example of a(n) _____________________ .

5.   The _____________________ of an equation is a drawing that represents all its solutions.

Objective a   Plot points associated with ordered pairs of numbers; determine the quadrant in which a point lies.

6. Plot these points. $(2, 4)$ $(-3, 1)$ $(2, -5)$ $(-4, -1)$ $(0, -2)$ **6.** ______________

$(2, 0)$ $(0, 3)$ $(-4, 0)$

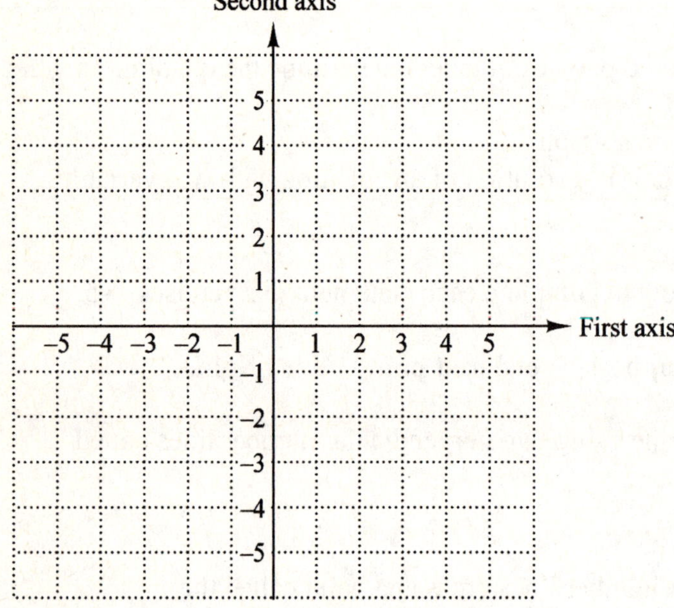

In which quadrant, if any, is the point located?

7. $(-2, 5)$ **7.** ______________

8. $(4, -10)$ **8.** ______________

9. $(0, 8)$ **9.** ______________

In which quadrant(s) can the point described be located?

10. The first coordinate is negative.   **10.** ______________

11. The first coordinate is the reciprocal of the second coordinate.   **11.** ______________

62

Objective b   Find the coordinates of a point on a graph.

Find the coordinates of the points A, B, C, D, and E.

12.

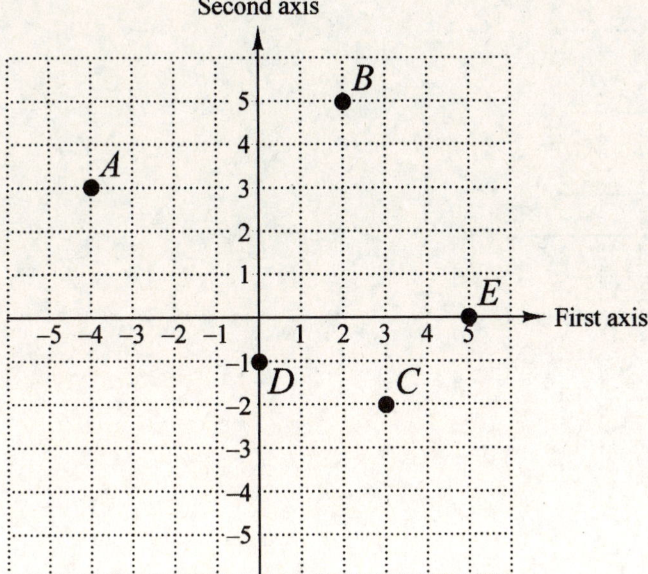

12. ______________

Objective c   Determine whether an ordered pair is a solution of an equation with two variables.

Determine whether the given ordered pair is a solution of the equation.

13. $(1, 5)$; $2x - y = 3$

13. ______________

14. $(-2, 4)$; $3a + 2b = 2$

14. ______________

15. $(0, -3)$; $5x - y = 3$

15. ______________

In Exercises 17 and 18, an equation and two ordered pairs are given. Show that each pair is a solution of the equation. Then use the graph of the two points to determine another solution. Answers may vary.

16. $y = 2x - 3;$ $(1, -1)$ and $(4, 5)$

16. ______________

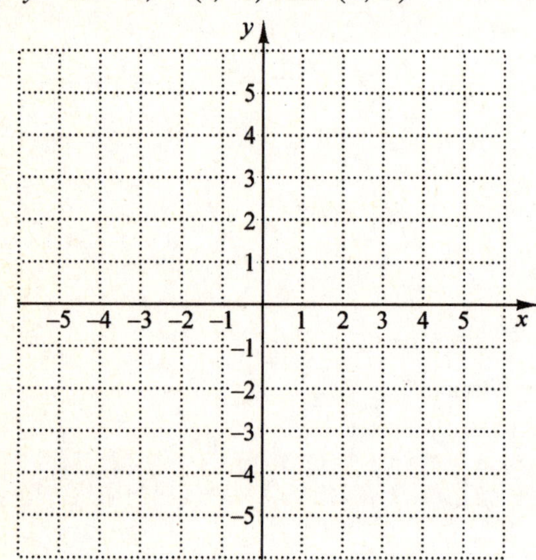

17. $x + 3y = 5;$ $(2, 1)$ and $(5, 0)$

17. ______________

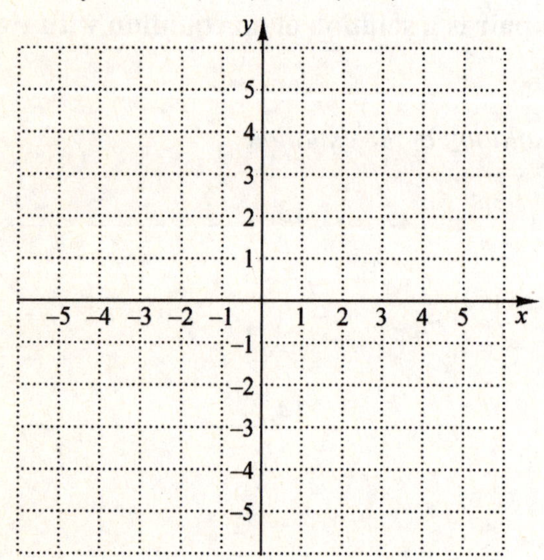

64

Chapter 3 GRAPHS OF LINEAR EQUATIONS

3.2 Graphing Linear Equations

Learning Objectives
a   Graph linear equations of the type $y = mx + b$ and $Ax + By = C$, identifying the y-intercept.
b   Solve applied problems involving graphs of linear equations.

Objective d   Graph linear equations of the type $y = mx + b$ and $Ax + By = C$, identifying the y-intercept.

Graph each equation and identify the y-intercept.

1. $y = \dfrac{2}{3}x$

x	y

1. _______________

2. $y = x - 2$

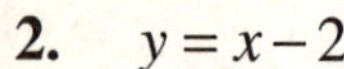

x	y

2. _______________

3. $x + y = 3$

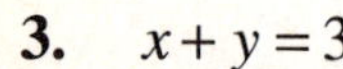

x	y

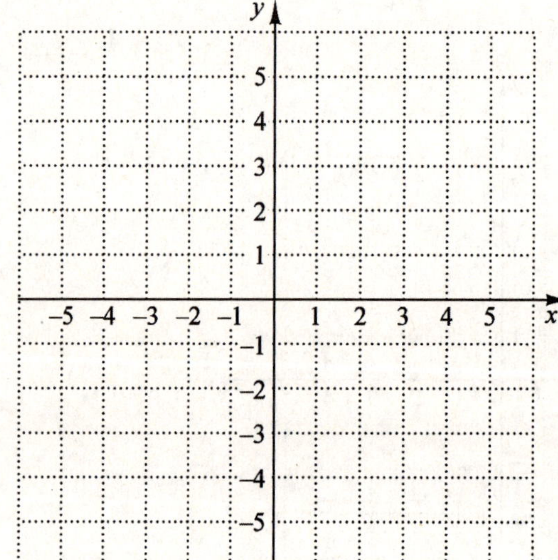

3. _______________

4. $y = -\dfrac{3}{2}x + 2$

x	y

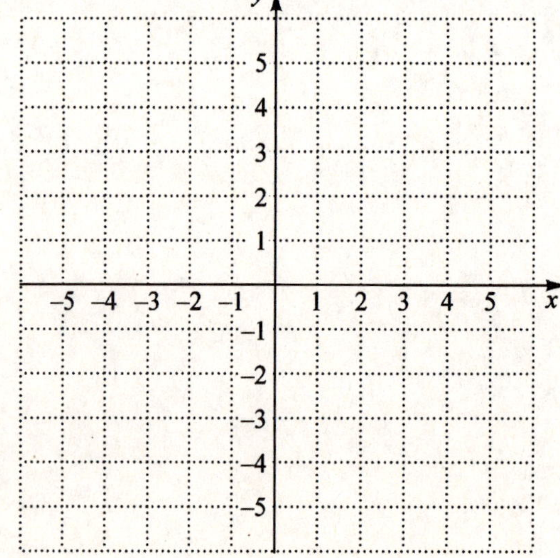

4. _______________

66

5. $x + 4y = 4$

x	*y*

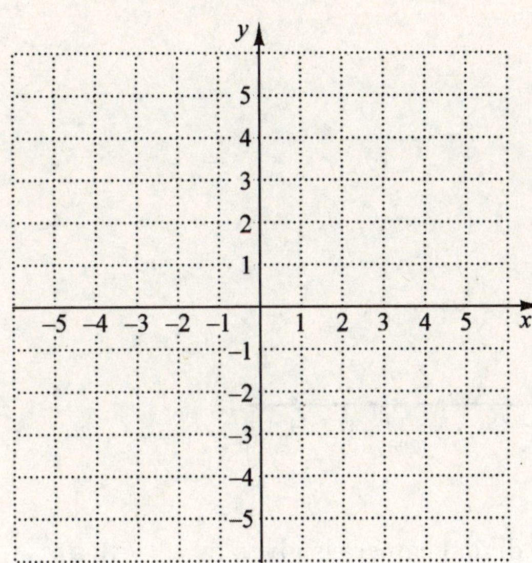

5. _______________

6. $5x - 2y = 10$

x	*y*

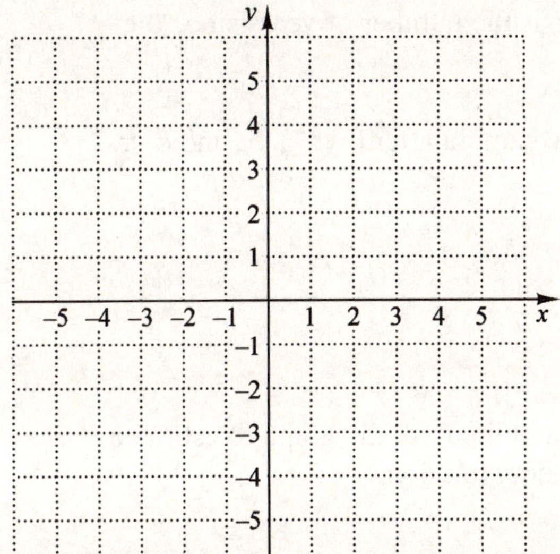

6. _______________

Objective b   Solve applied problems involving graphs of linear equations.

Solve.

7. The average annual expenditure on home insurance can be approximated by $H = 21n + 414$, where n is the number of years since 1995.

Sources: National Association of Insurance Commissioners, Insurance firms

   a)   Find the average annual home insurance cost in 1996 $(n = 1)$, 1998, 2004, and 2007.

7. a) _______________

b) _______________

67

b) Graph the equation and use the graph to estimate
   what the average annual home insurance expenditure
   was in 2001.

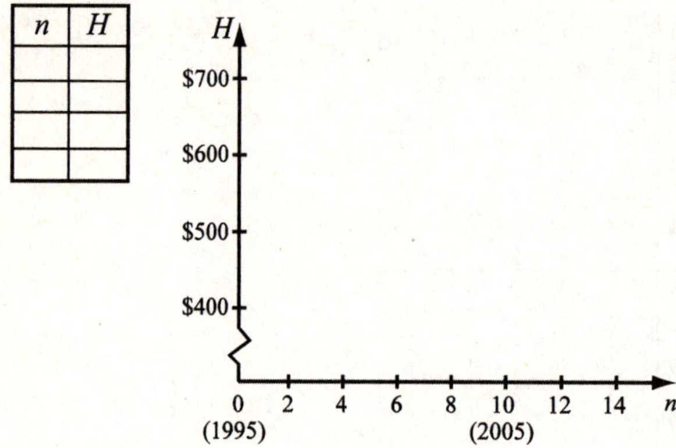

8. The value V, in dollars, of a digital camera is given by
   $V = -75t + 450$, where t is the number of years since the
   camera was purchased.

   a) Find the value of the camera after 0 yr, 3 yr, and 6 yr.

   b) Graph the equation and then use the graph to estimate
      the value of the camera after 4 yr.

8. a)_____________________

   b)_____________________

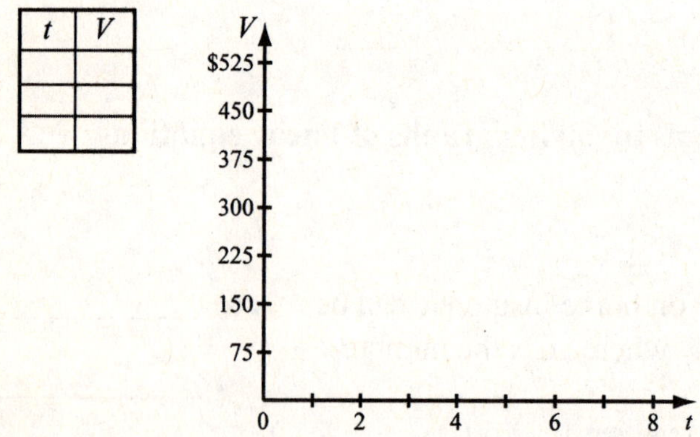

68

Chapter 3 GRAPHS OF LINEAR EQUATIONS

3.3 More with Graphing and Intercepts

Learning Objectives
a   Find the intercepts of a linear equation, and graph using intercepts.
b   Graph equations equivalent to those of the type $x = a$ and $y = b$.

Key Terms
Use the vocabulary terms listed below to complete each statement in Exercises 1–4.

  horizontal line   **vertical line**   *x*-intercept   *y*-intercept

1. The ______________________ occurs where a graph crosses the *x*-axis.

2. The ______________________ occurs where a graph crosses the *y*-axis.

3. The graph of $x = a$ is a(n) ______________________.

4. The graph of $y = b$ is a(n) ______________________.

Objective a   Find the intercepts of a linear equation, and graph using intercepts.

For Exercises 5 and 6, find (a) the coordinates of the y-intercept and (b) the coordinates of the x-intercept.

5.

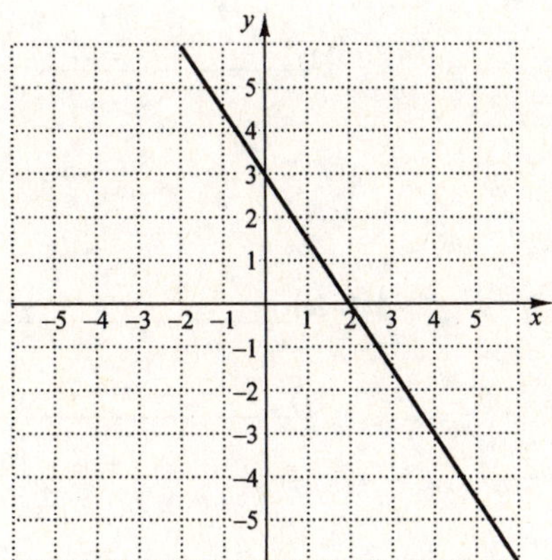

5. a)_____________________

   b)_____________________

6.

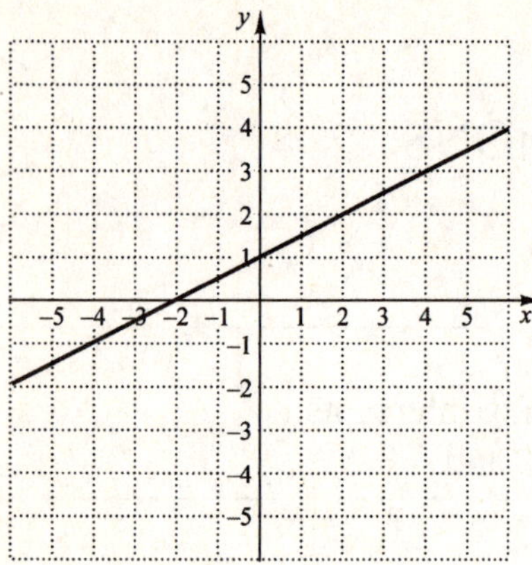

6. a)_______________

b)_______________

For Exercises 7–10, find (a) the coordinates of the y-intercept and (b) the coordinates of the x-intercept. Do not graph.

7. $6x + 5y = 30$

7. a)_______________

b)_______________

8. $2x - 5y = 40$

8. a)_______________

b)_______________

9. $-3x + 2y = 5$

9. a)_______________

b)_______________

10. $4x - 2 = 6y$

10. a)_______________

b)_______________

70

For each equation, find the intercepts. Then use the intercepts to graph the equation.

11. $x + 2y = 4$ **11.** _______________

x	y
0		← y-intercept
	0	← x-intercept

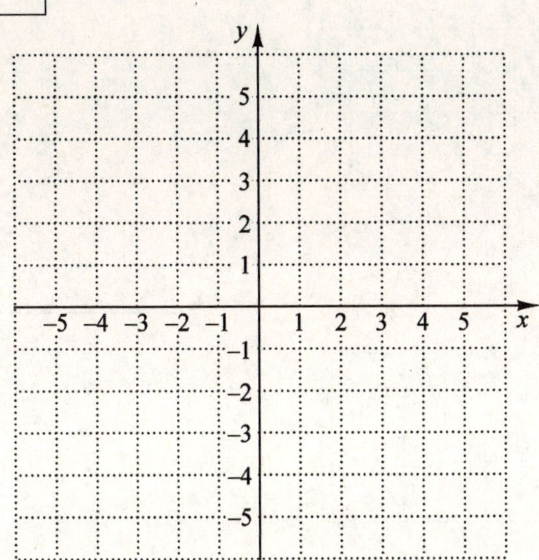

12. $3y - 3 = x$ **12.** _______________

x	y
0		← y-intercept
	0	← x-intercept

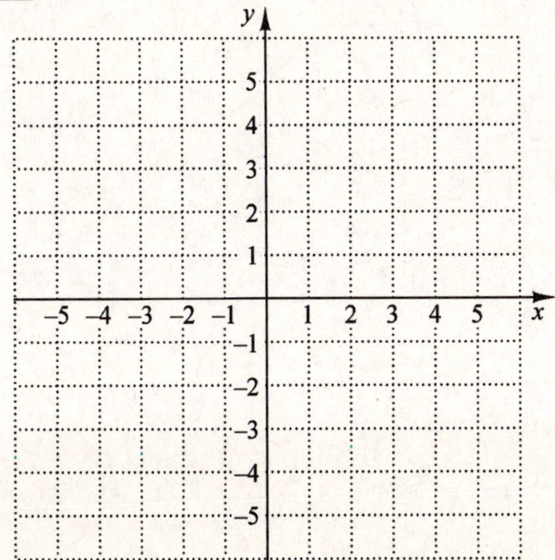

71

13. $6x - 12 = 6y$

x	y

13. _______________

14. $x + 5 = y$

x	y

14. _______________

72

15. $4x - 5y = 20$

x	y

15. _______________

16. $y + 5x = 0$

x	y

16. _______________

Objective b   Graph equations equivalent to those of the type $x = a$ and $y = b$.

Graph.

17. $x = -3$

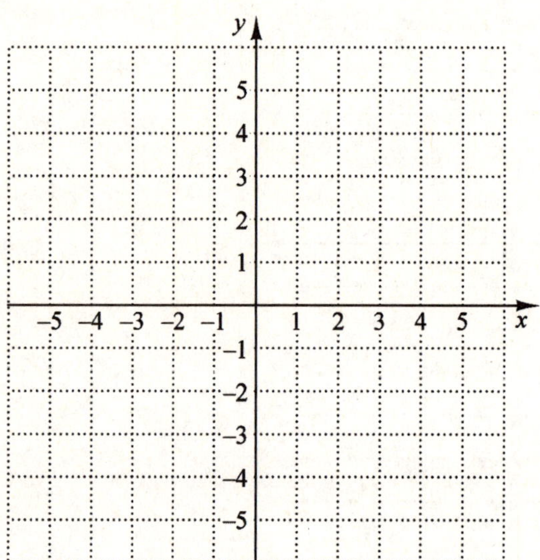

17. _________________

18. $y = 4$

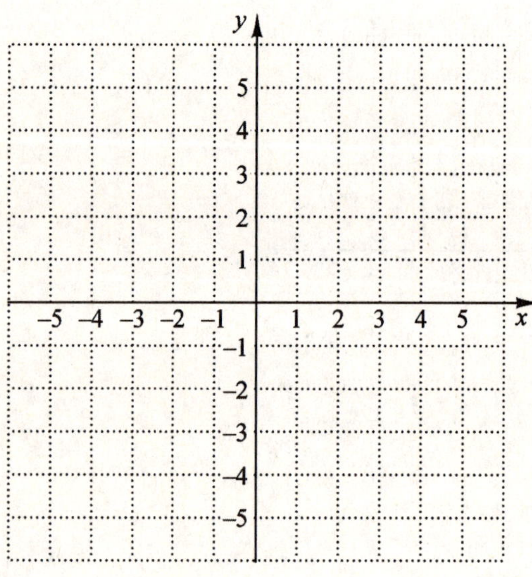

18. _________________

74

19. $2y = -5$

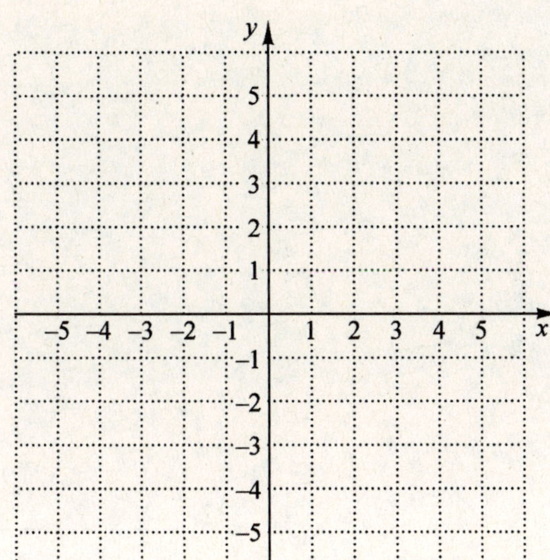

20. $3x - 15 = 0$

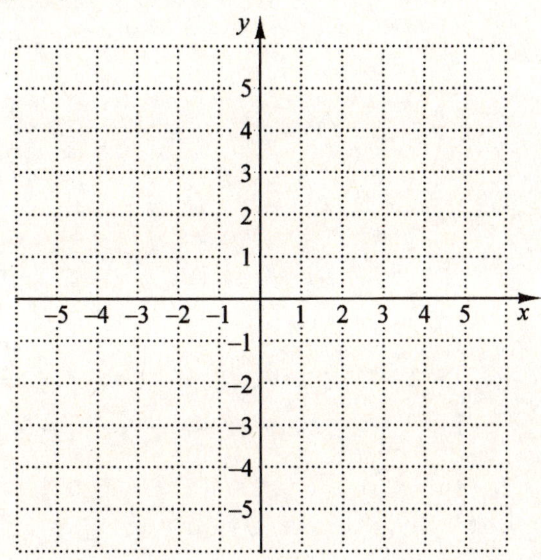

21.

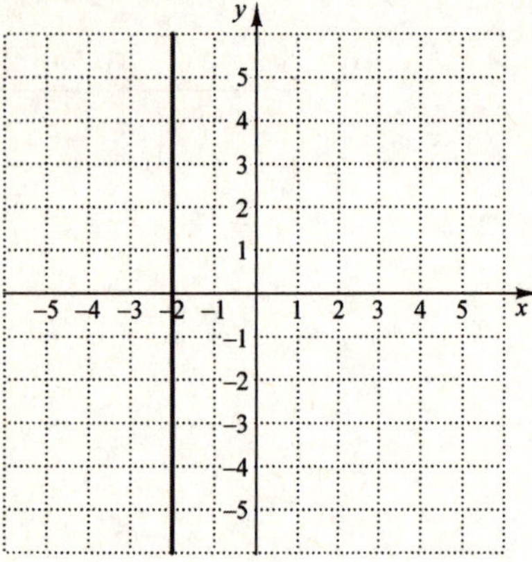

21. _______________

22.

22. _______________

76

Chapter 3 GRAPHS OF LINEAR EQUATIONS

3.4 Slope and Applications

<table>
<tr><td>Learning Objectives</td></tr>
<tr><td>a   Given the coordinates of two points on a line, find the slope of the line, if it exists.
b   Find the slope of a line from an equation.
c   Find the slope, or rate of change, in an applied problem involving slope.</td></tr>
</table>

Key Terms

Use the vocabulary terms listed below to complete each statement in Exercises 1–4.

grade   **rise**   **run**   **slope**

1. The change in y as we move from one point to another is called the

   _________________ .

2. The change in x as we move from one point to another is called the

   _________________ .

3. The _________________ of a line is the ratio of the change in y to the change in x.

4. The _________________ of the road is a measure of how steep it is.

Objective a   Given the coordinates of two points on a line, find the slope of the line, if it exists.

Find the slope, if it exists, of the line.

5.

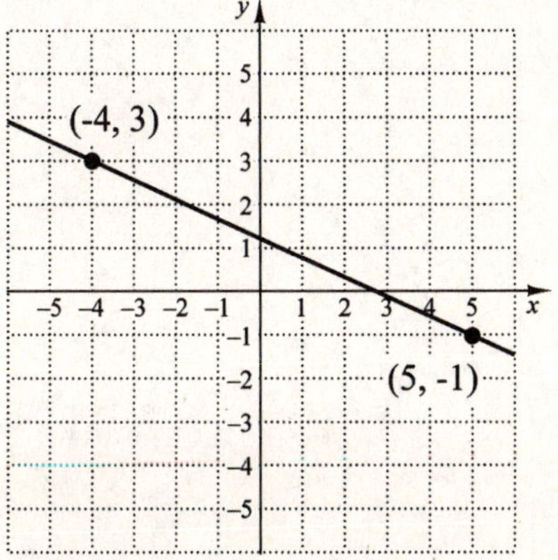

5. _________________

6.

y
5
4
3
2
1
-5 -4 -3 -2 -1   1 2 3 4 5   x
-1
-2
-3
-4
-5

6. _______________

7.

y
5
(-3, 4)   4
3
2
1
-5 -4 -3 -2 -1   1 2 3 4 5   x
(-3, -1)   -1
-2
-3
-4
-5

7. _______________

8.

y
(-2, 4)   5   (5, 4)
4
3
2
1
-5 -4 -3 -2 -1   1 2 3 4 5   x
-1
-2
-3
-4
-5

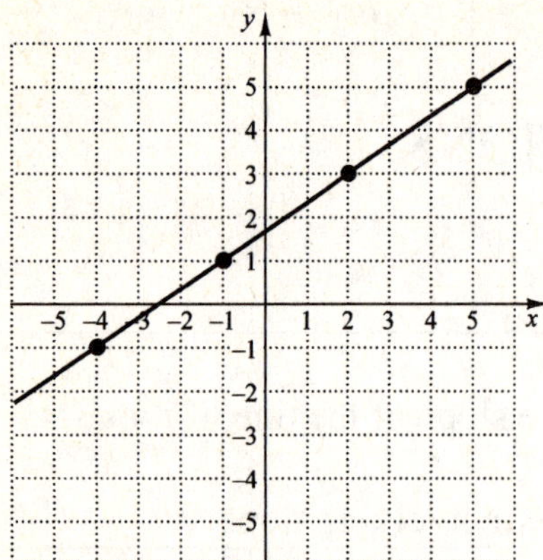

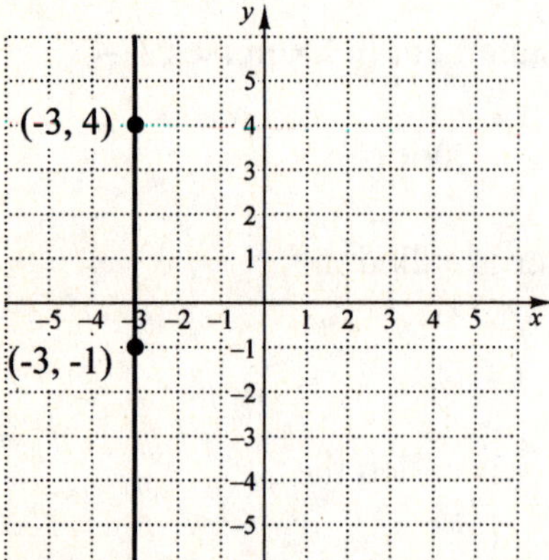

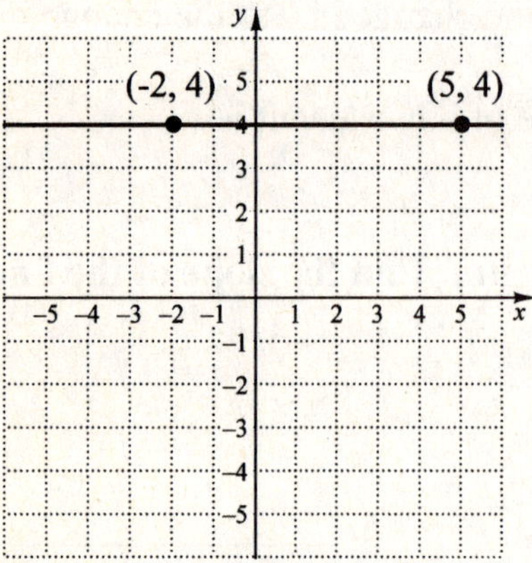

8. _______________

78

Graph the line containing the given pair of points and find the slope.

9. $(-4, 2),\ (1, 0)$

9. ________________

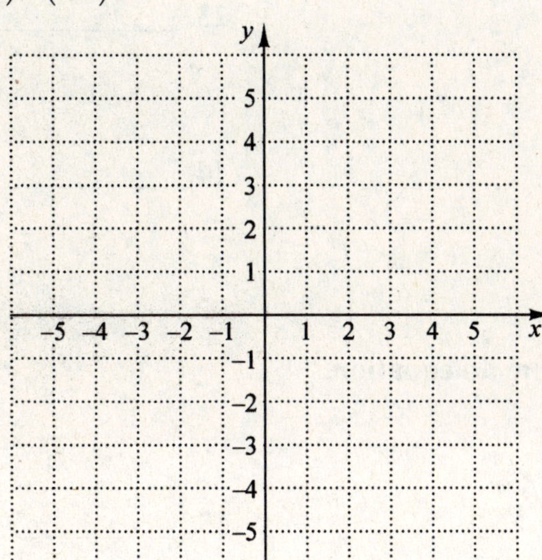

10. $(-2, -5),\ (4, 1)$

10. ________________

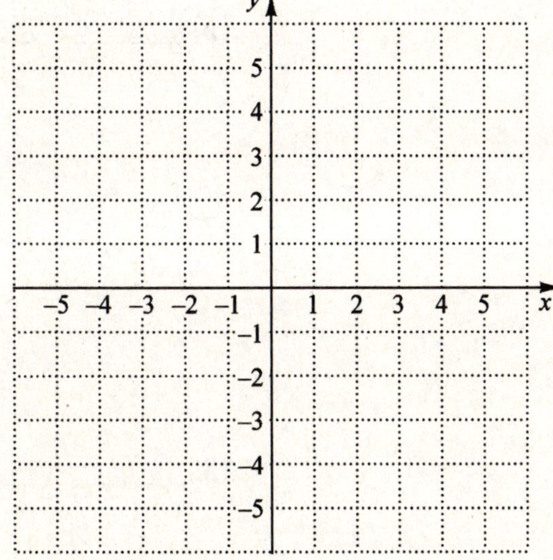

Find the slope, if it exists, of the line containing the given pair of points.

11. $\left(3, -\dfrac{1}{3}\right),\ \left(-2, \dfrac{5}{4}\right)$

11. ________________

12. $\left(\frac{1}{2},-2\right), \left(\frac{1}{2},-5\right)$

12. _______________

13. $(0.6,-0.8), (0.1,-0.8)$

13. _______________

14. $(10,\ 15), (-25,-8)$

14. _______________

Objective b **Find the slope of a line from an equation.**

Find the slope, if it exists, of the line.

15. $y = -6x + 8$

15. _______________

16. $5x - y = 2$

16. _______________

17. $x = \frac{5}{9}$

17. _______________

18. $x + 3y = 6$

18. _______________

19. $4x + 8y = 16$

19. _______________

80

20. $y = 18$ 20. _______________

21. $y = 13.8x$ 21. _______________

22. $10x = 2y - 7$ 22. _______________

Objective c   Find the slope, or rate of change, in an applied problem involving slope.

23. A road rises 7 ft over a horizontal distance of 280 ft. Find   23. _______________
the slope (or grade) of the road.

24. A ramp rises 3 ft over a horizontal distance of 50 ft. Find   24. _______________
the grade of the ramp.

In Exercises 25 and 26, use the graph to calculate the rate of change in which the units of the horizontal axis are used in the denominator.

25. The following graph shows data for a car driven in the city. Find the rate of change in miles per gallon, that is, the gas mileage.

25. _________________

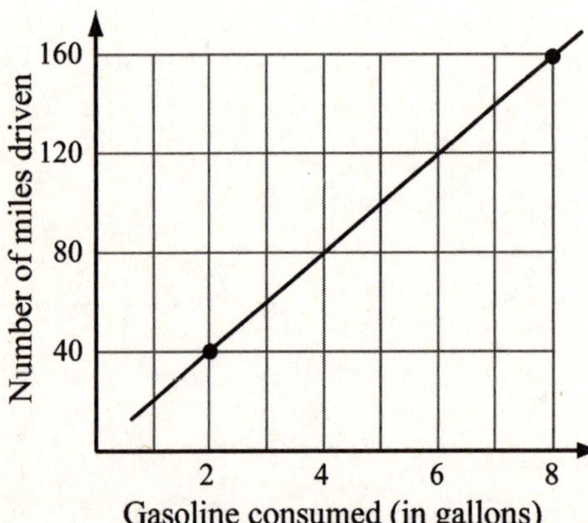

26. The per capita consumption in the United States of whole milk is represented in the following graph. Find the rate of change in consumption with respect to time, in gallons per year.
Source: US Department of Agriculture

26. _________________

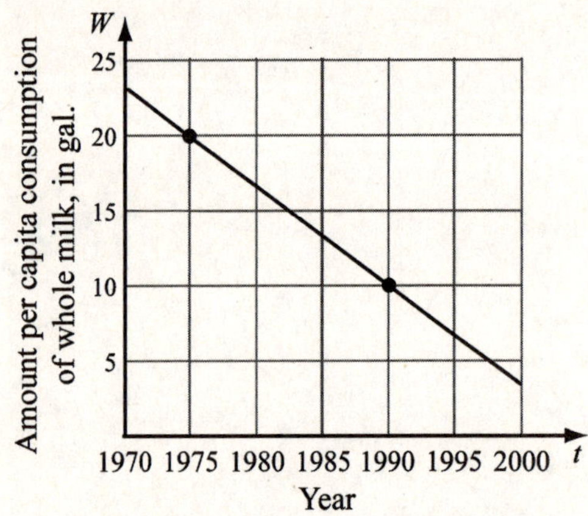

82

Chapter 4 POLYNOMIALS: OPERATIONS

4.1 Integers as Exponents

Learning Objectives
a   Tell the meaning of exponential notation.
b   Evaluate exponential expressions with exponents of 0 and 1.
c   Evaluate algebraic expressions containing exponents.
d   Use the product rule to multiply exponential expressions with like bases.
e   Use the quotient rule to divide exponential expressions with like bases.
f   Express an exponential expression involving negative exponents with positive exponents.

Key Terms
Use the vocabulary terms listed below to complete each statement in Exercises 1–4.

   base   **exponent**   ***x*-cubed**   ***x*-squared**

1.   In the expression 3^7, 7 is the ___________________.

2.   In the expression 3^7, 3 is the ___________________.

3.   We often read x^2 as "___________________."

4.   We often read x^3 as "___________________."

Objective a   Tell the meaning of exponential notation.

What is the meaning of each of the following?

5. 7^3 5. _______________

6. $\left(\dfrac{3}{4}\right)^5$ 6. _______________

7. $(8x)^4$ 7. _______________

8. $-9y^2$ 8. _______________

Objective b   Evaluate exponential expressions with exponents of 0 and 1.

Evaluate.

9. $\left(\dfrac{3}{4}\right)^{0}$

9. _______________

10. 5.17^{0}

10. _______________

11. $(mn)^{0}$

11. _______________

12. mn^{0}

12. _______________

Objective c   Evaluate algebraic expressions containing exponents.

Evaluate.

13. x^{5}, when $x = 2$

13. _______________

14. $-t^{2}$, when $t = -5$

14. _______________

15. $a^{2} - 3$, when $a = -4$

15. _______________

16. $x^{1} + 10$ and $x^{0} + 10$, when $x = 12$

16. _______________

Objective d   Use the product rule to multiply exponential expressions with like bases.

Multiply and simplify.

17. $7^{4} \cdot 7^{6}$

17. _______________

18. $(4p)^{5}(4p)^{3}$

18. _______________

84

19. $x^{11} \cdot x^{10}$ **19.** _______________

20. $n^0 \cdot n^9$ **20.** _______________

Objective e   Use the quotient rule to divide exponential expressions with like bases.

Divide and simplify.

21. $\dfrac{11^8}{11^3}$ **21.** _______________

22. $\dfrac{x^6}{x^2}$ **22.** _______________

23. $\dfrac{t^{10}}{t}$ **23.** _______________

24. $\dfrac{(3p)^5}{(3p)^5}$ **24.** _______________

Objective f   Express an exponential expression involving negative exponents with positive exponents.

Express using positive exponents. Then simplify.

25. 4^{-2} **25.** _______________

26. 10^{-2} **26.** _______________

27. $\dfrac{1}{x^{-5}}$

27. _______________

28. $\dfrac{1}{x^{-n}}$

28. _______________

Express using negative exponents.

29. $\dfrac{1}{x^4}$

29. _______________

30. $\dfrac{1}{a^6}$

30. _______________

Multiply and simplify.

31. $5^2 \cdot 5^{-4}$

31. _______________

32. $x^{-2} \cdot x^{-1}$

32. _______________

33. $c^8 \cdot c^{-4} \cdot c^{-5}$

33. _______________

34. $y^7 \cdot y^{-7}$

34. _______________

Divide and simplify.

35. $\dfrac{7^5}{7^6}$

35. _______________

36. $\dfrac{x}{x^{-3}}$

36. _______________

37. $\dfrac{t^{-5}}{t^{-1}}$

37. _______________

38. $\dfrac{3^{-6}}{3^{-6}}$

38. _______________

86

Chapter 4 POLYNOMIALS: OPERATIONS

4.2 Exponents and Scientific Notation

Learning Objectives
a   Use the power rule to raise powers to powers.
b   Raise a product to a power and a quotient to a power.
c   Convert between scientific notation and decimal notation.
d   Multiply and divide using scientific notation.
e   Solve applied problems using scientific notation.

Key Terms
Use the vocabulary terms listed below to complete each statement in Exercises 1–2.

    exponential notation   **scientific notation**

1.   The expression 4^5 is written in __________________.

2.   The expression 1.3×10^8 is written in __________________.

Objective a   **Use the power rule to raise powers to powers.**

Objective b   **Raise a product to a power and a quotient to a power.**

Simplify.

3. $\left(3^2\right)^4$ 3. ______________

4. $\left(a^5\right)^{-3}$ 4. ______________

5. $\left(x^{-2}\right)^{-7}$ 5. ______________

6. $\left(t^8\right)^{-5}$ 6. ______________

7. $\left(xy^2\right)^{-3}$ 7. ______________

8. $\left(m^2 n^{-5}\right)^{-4}$ 8. ______________

9. $\left(a^2b^{-3}c^{-6}\right)^{-2}$

10. $\left(-5x^4y^{-3}\right)^2$

11. $\left(\dfrac{x^5}{3}\right)^2$

12. $\left(\dfrac{a^4}{5}\right)^{-2}$

13. $\left(\dfrac{p^2q^4}{n}\right)^5$

14. $\left(\dfrac{xy^2}{w^3z}\right)^{-3}$

9. _______________

10. _______________

11. _______________

12. _______________

13. _______________

14. _______________

Objective c   Convert between scientific notation and decimal notation.

Convert to scientific notation.

15. 460,000,000,000

16. 203,100,000,000,000

17. 0.000000405

18. 0.00000000027

15. _______________

16. _______________

17. _______________

18. _______________

Convert to decimal notation.

19. 4.02×10^8

20. 3.27×10^{-5}

19. _______________

20. _______________

88

21. 10^{-4}

22. 10^{5}

Objective d   Multiply and divide using scientific notation.

Multiply or divide and write scientific notation for the result.

23. $\left(2.1\times10^{5}\right)\left(3.5\times10^{-9}\right)$

24. $\left(4.3\times10^{-2}\right)\left(2.4\times10^{-8}\right)$

25. $\dfrac{4.2\times10^{-3}}{1.25\times10^{7}}$

26. $\dfrac{8.5\times10^{8}}{1.7\times10^{-11}}$

21. _________________

22. _________________

23. _________________

24. _________________

25. _________________

26. _________________

Objective e   Solve applied problems using scientific notation.

Solve.

27. Light travels approximately 3×10^8 m/s. How far does
    light travel in 1 year?

27. _______________

28. Red blood cells measure approximately 6×10^{-6} m in
    diameter. At the point at which the aorta begins at the
    upper part of the left ventricle, its diameter is
    approximately 3×10^{-2} m. How many times larger is the
    diameter of the aorta at that point than the diameter of a
    red blood cell?

28. _______________

90

Chapter 4 POLYNOMIALS: OPERATIONS

4.3 Introduction to Polynomials

Learning Objectives

a   Evaluate a polynomial for a given value of the variable.
b   Identify the terms of a polynomial.
c   Identify the like terms of a polynomial.
d   Identify the coefficients of a polynomial.
e   Collect the like terms of a polynomial.
f   Arrange a polynomial in descending order, or collect the like terms and then arrange in descending order.
g   Identify the degree of each term of a polynomial and the degree of the polynomial.
h   Identify the missing terms of a polynomial.
i   Classify a polynomial as a monomial, binomial, trinomial, or none of these.

Key Terms

Use the vocabulary terms listed below to complete each statement in Exercises 1–8.

binomial	coefficient	degree	descending order
like terms	monomial	trinomial	value

1.   A polynomial with just one term is a(n) _____________________ .

2.   When we replace the variable in a polynomial with a number, the polynomial represents

       a(n) _____________________ of the polynomial.

3.   Terms that have the same variable raised to the same power are _____________________ .

4.   The _____________________ of $4x^3$ is 4.

5.   The _____________________ of $4x^3$ is 3.

6.   When a polynomial is arranged in _____________________, the exponents decrease from

       left to right.

7.   A polynomial with two terms is a(n) _____________________.

8.   A polynomial with three terms is a(n) _____________________.

Objective a   Evaluate a polynomial for a given value of the variable.

Evaluate the polynomial when $x = -3$ and when $x = 2$.

9. $x^2 - 3x + 1$

9. _________________

10. $-2x^3 + x^2 + 3x - 5$

10. _________________

11. The graph of the polynomial equation $y = x^2 - 3$ is shown below. Use *only* the graph to estimate the value of the polynomial when $x = -2$, $x = -1$, $x = 0$, $x = 1.5$, and $x = 3$.

11. _________________

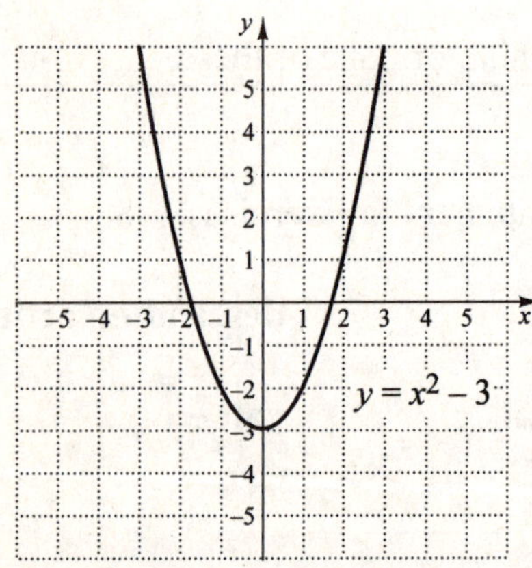

12. The percent growth r in U.S. metropolitan apartment rent can be approximated by the polynomial equation $r = -0.03x^4 + 0.589x^3 - 3.426x^2 + 5.557x + 2.327$, where x is the number of years since 1999. Use the graph to estimate the percent growth in 2000 and in 2005.

12. _________________

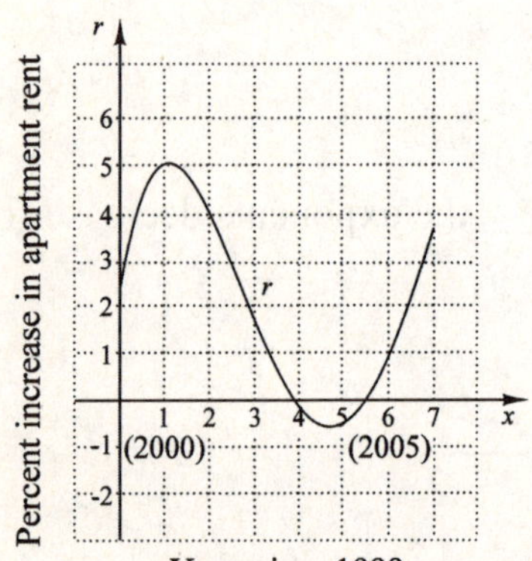

92

Objective b   Identify the terms of a polynomial.

Identify the terms of each polynomial.

13. $5 - 4x^2 + x$

13. _________________

14. $-5x^6 + \dfrac{2}{3}x^4 - 3x + 7$

14. _________________

Objective c   Identify the like terms of a polynomial.

Identify the like terms in each polynomial.

15. $4x^3 - 6x^2 + 2x^3 - x^2$

15. _________________

16. $2x^4 + 3x - 7 + x^4 - 5x + 11$

16. _________________

Objective d   Identify the coefficients of a polynomial.

Identify the coefficient of each term of the polynomial.

17. $3x^2 + \dfrac{1}{4}x - 7$

17. _________________

18. $-4x^3 - 5x^2 + 1.3x + 8$

18. _________________

93

Objective e   Collect the like terms of a polynomial.

Collect like terms.

19. $x - 10x$ 19. _______________

20. $2x^3 + 3x - x^3 - 10x$ 20. _______________

21. $\dfrac{1}{2}x^4 + 7 + \dfrac{3}{4}x^4 - x^2 - 18$ 21. _______________

22. $10x^2 - 8x^3 + 3x^2 + 11x^3 - 12x^2$ 22. _______________

Objective f   Arrange a polynomial in descending order, or collect the like terms and then arrange in descending order.

Arrange the polynomial in descending order.

23. $2x^2 - x + 5x^3 - 6 + x^5$ 23. _______________

24. $12y^3 + y^7 - y + 6 + 18y^2$ 24. _______________

Collect like terms and then arrange in descending order.

25. $-x - 5x^2 - 8x + 14x^2 + 11$ 25. _______________

26. $3x - \dfrac{2}{3} + 5x^5 - \dfrac{1}{3} + 8x^5 - 15x$ 26. _______________

94

Objective g   Identify the degree of each term of a polynomial and the degree of the polynomial.

Identify the degree of each term of the polynomial and the degree of the polynomial.

27. $6x^4 - 5x + 7$

27. ______________

28. $3x + 1 + 8x^3 - 4x^6$

28. ______________

Objective h   Identify the missing terms of a polynomial.

Identify the missing terms in each polynomial.

29. $x^6 - x^2 + x$

29. ______________

30. $3x^4 - 2x^3 - x - 4$

30. ______________

Write the polynomial in two ways: with its missing terms and by leaving space for them.

31. $x^3 - 8$

31. ______________

32. $4x^5 - x^3 + x^2 - 1$

32. ______________

Classify each polynomial as a monomial, binomial, trinomial, or none of these.

33. $-3x^2$

34. $2x^2 - 8x + 5$

35. $4x^3 + 6x^2 + 6x + 4$

36. $x^2 - 1$

33. _______________

34. _______________

35. _______________

36. _______________

96

Chapter 4 POLYNOMIALS: OPERATIONS

4.4 Addition and Subtraction of Polynomials

Learning Objectives
a   Add polynomials.
b   Simplify the opposite of a polynomial.
c   Subtract polynomials.
d   Use polynomials to represent perimeter and area.

Key Terms
Use the vocabulary terms listed below to complete each statement in Exercises 1–2.

opposite   **sign**

1.   To find the additive inverse of a polynomial, change the _____________________ of every term.

2.   To subtract polynomials, add the _____________________ of the polynomial being subtracted.

Objective a   Add polynomials.

Add.

3. $(2x-7)+(-3x+2)$ 3. _______________

4. $(-x+5)+\left(2x^2+\dfrac{1}{3}x-10\right)$ 4. _______________

5. $(6x^2-4x+8)+(9x^2+4x-12)$ 5. _______________

6. $\left(2.3x^3 + 1.6x^2 - 2.7x\right) + \left(-5.7x^3 - 3.2x^2 + 19\right)$

6. _______________

7. $\left(7 + 3x + 8x^2 + 9x^3\right) + \left(3 - 9x - 8x^2 - 9x^3\right)$

7. _______________

8. $\left(\dfrac{2}{5}x^4 + \dfrac{3}{8}x^3 - \dfrac{3}{10}x^2 + 16\right) + \left(-\dfrac{7}{5}x^4 - \dfrac{9}{10}x^2 - 20\right)$

8. _______________

9. $\left(3x^7 - 4x^5 + 2x^3 - x\right) + \left(8x^6 - 12x^5 + 3\right) + \left(-x^5 + 8x^3 + 5x\right)$

9. _______________

10.

$$
\begin{array}{rrrrr}
0.8x^4 - 0.15x^3 & - 0.01x^2 & & + 0.1 \\
- 0.25x^3 & - 0.01x^2 & + 2x & \\
0.7x^4 & + 0.05x^2 & & - 0.5 \\
0.40x^3 & & & + 0.9 \\
- 0.6x^4 & + 0.03x^2 & + 15x & - 2 \\
\hline
\end{array}
$$

10. _______________

Objective b   Simplify the opposite of a polynomial.

Simplify.

11. $-(-10y)$

11. _______________

12. $-\left(-a^2 - 4a + \dfrac{5}{2}\right)$

12. _______________

98

13. $-\left(6x^5 - 7x + 10\right)$ 13. ______________

14. $-\left(-3x + 8\right)$ 14. ______________

15. $-\left(-5x^2 - 6x + 8\right)$ 15. ______________

16. $-\left(9x^4 + 3x^2 - \dfrac{1}{3}x - 1\right)$ 16. ______________

Objective c   Subtract polynomials.

Subtract.

17. $\left(-2x - 3\right) - \left(x^2 + 5x - 5\right)$ 17. ______________

18. $\left(y^2 - 7\right) - \left(y^2 + 7\right)$ 18. ______________

19. $\left(8x^4 + 2x^3 - 5\right) - \left(7x^3 - x^2 - 1\right)$ 19. ______________

20. $\left(1.3x^3 + 2.4x^2 - 1.7x\right) - \left(-4.7x^3 - 3.8x^2 + 5.6\right)$ 20. ______________

99

21. $\left(\dfrac{2}{3}x^3 - \dfrac{9}{10}x - \dfrac{5}{4}\right) - \left(-\dfrac{1}{3}x^3 + \dfrac{9}{10}x - \dfrac{5}{4}\right)$

21. _________________

22.
$$\begin{array}{r} x^2 + 6x + 9 \\ -\left(x^2 + 5x\right) \\ \hline \end{array}$$

22. _________________

23.
$$\begin{array}{r} 4x^4 + 8x^3 - x^2 \\ -\left(-2x^4 - 8x^3 \quad\quad + 7x + 11\right) \\ \hline \end{array}$$

23. _________________

24.
$$\begin{array}{r} x^4 \quad\quad\quad\quad -1 \\ -\left(x^4 + x^3 + x^2 - x - 1\right) \\ \hline \end{array}$$

24. _________________

Objective d   Use polynomials to represent perimeter and area.

Solve.

25. Find a polynomial for the perimeter of the figure.

25. _________________

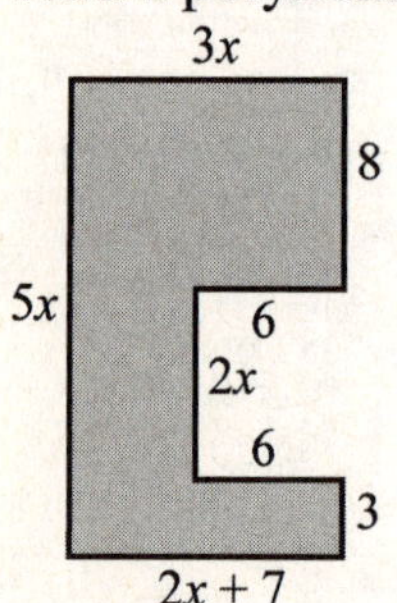

100

Name: Date:
Instructor: Section:

26. Find a polynomial for the sum of the areas of these
rectangles.

26. _______________

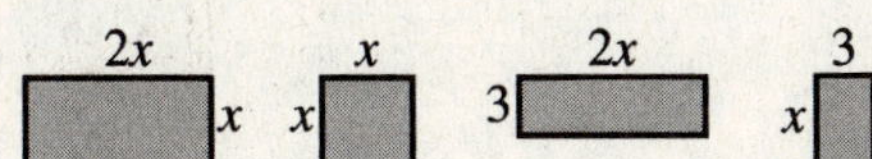

*Find two algebraic expressions for the area of each figure. First, regard the figure as
one large rectangle, and then regard the figure as a sum of four smaller rectangles.*

27.

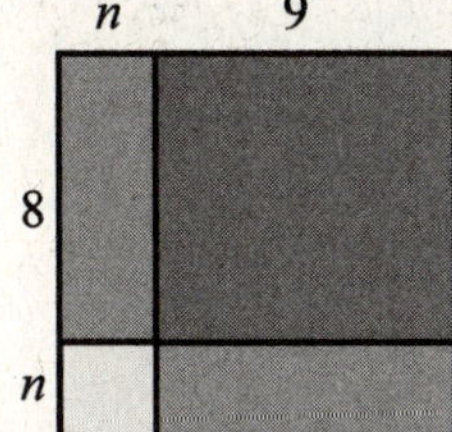

27. _______________

28.

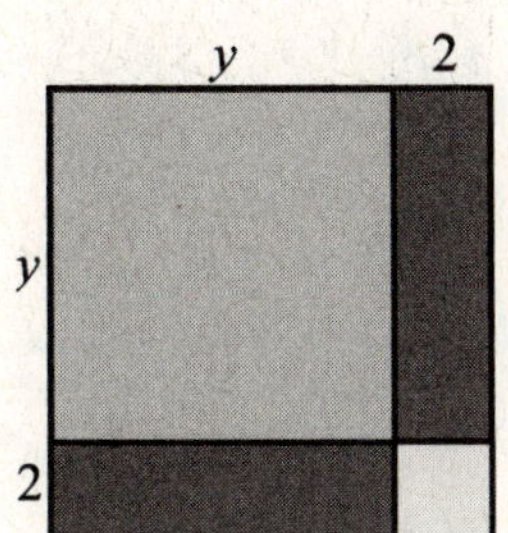

28. _______________

Find a polynomial for the shaded area of the figure.

29.

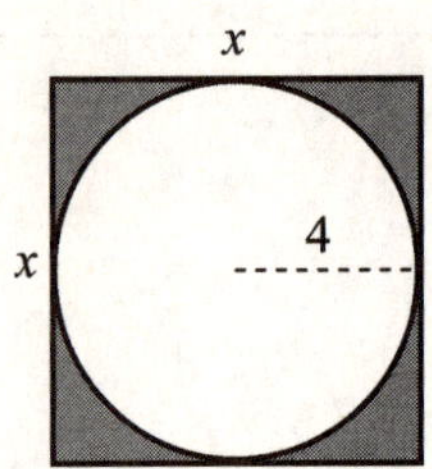

29. ____________________

30.

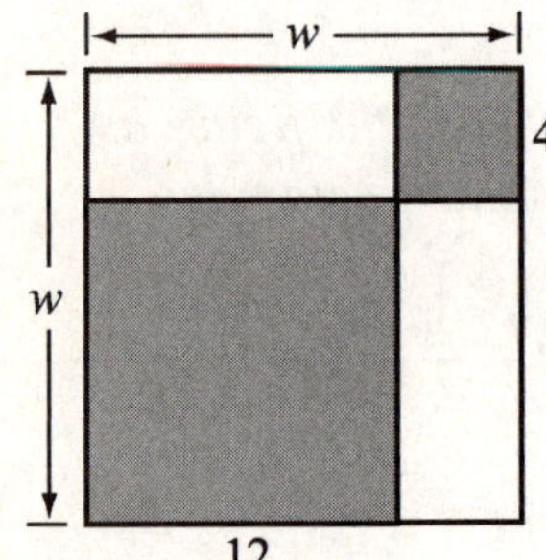

30. ____________________

102

Chapter 4 POLYNOMIALS: OPERATIONS

4.5 Multiplication of Polynomials

Learning Objectives
a   Multiply monomials.
b   Multiply a monomial and any polynomial.
c   Multiply two binomials.
d   Multiply any two polynomials.

Key Terms
Use the vocabulary terms listed below to complete each statement in Exercises 1–2.

     collect   **term**

1.   To multiply a monomial and a polynomial, multiply each ___________________ of the polynomial by the monomial.

2.   After multiplying, if possible, ___________________ like terms.

Objective a   Multiply monomials.

Multiply.

3. $\left(3a^4\right)\left(7a^5\right)$ 3. ______________

4. $\left(-0.2x^3\right)\left(0.5x^7\right)$ 4. ______________

5. $\left(-\dfrac{1}{4}x\right)\left(-\dfrac{1}{5}x^8\right)$ 5. ______________

6. $\left(-3t^4\right)\left(5t^2\right)\left(-2t^5\right)$ 6. ______________

Objective b   Multiply a monomial and any polynomial.

Multiply.

7. $-4x(2x-3)$

7. ________________

8. $10x(5x^2-8x+7)$

8. ________________

9. $(6y^3)(2y^5+5y^8)$

9. ________________

10. $(-5a^2)(a^4+a)$

10. ________________

Objective c   Multiply two binomials.

Multiply.

11. $(x+1)(x+8)$

11. ________________

12. $(x+5)(x-7)$

12. ________________

13. $(2x-3)(2x-3)$

13. ________________

14. $(6-5x)(6-x)$

14. ________________

104

Name: Date:
Instructor: Section:

15. $\left(x-\frac{2}{3}\right)\left(x+\frac{3}{2}\right)$

15. ______________

16. $(x+1.4)(x-2.5)$

16. ______________

Write an algebraic expression that represents the total area of the four smaller rectangles.

17. 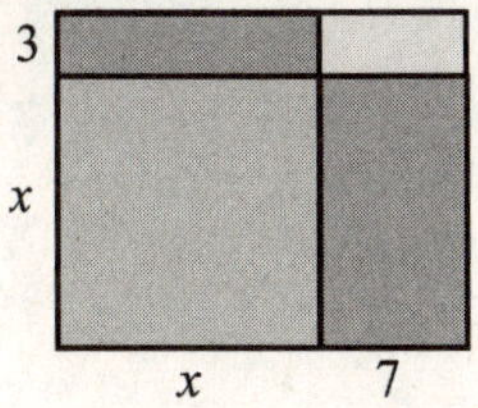

17. ______________

18. 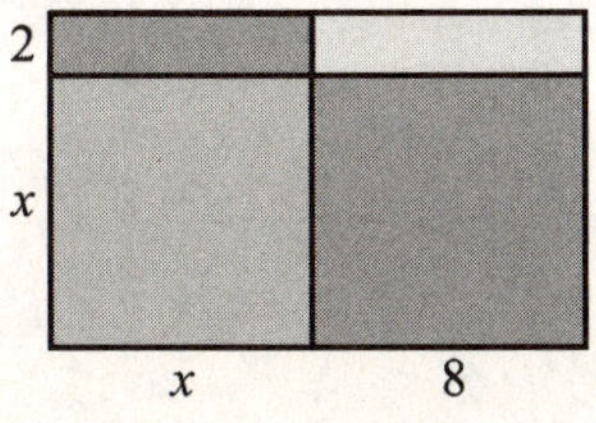

18. ______________

Draw and label rectangles to illustrate each product.

19. $x(x+3)$

19. _______________

20. $(x+4)(x+5)$

20. _______________

Objective d   Multiply any two polynomials.

Multiply.

21. $(x^2+x+2)(x-2)$

21. _______________

22. $(3x+2)(2x^2+x+5)$

22. _______________

23. $(x^2-4)(3x^2-2x+1)$

23. _______________

106

24. $\left(y^3 + y^2\right)\left(y^3 - 2y^2 + y\right)$ 24. ______________

25. $\left(-4x^3 - 8x^2 + 5\right)\left(3x^2 - x\right)$ 25. ______________

26. $\left(1 + a + a^2\right)\left(1 - a - a^2\right)$ 26. ______________

27. $\left(3n^2 - n - 5\right)\left(2n^2 + 5n - 3\right)$ 27. ______________

28. $\left(t - t^2 + t^4\right)\left(t^3 - 1 + t^5\right)$ 28. ______________

29. $(y+1)(y^3+4y^2-2y-7)$

29. _______________

30. $\left(x+\dfrac{1}{2}\right)\left(8x^3+5x^2+6x-\dfrac{1}{3}\right)$

30. _______________

108

Chapter 4 POLYNOMIALS: OPERATIONS

4.6 Special Products

Learning Objectives
a   Multiply two binomials mentally using the FOIL method.
b   Multiply the sum and the difference of two terms mentally.
c   Square a binomial mentally.
d   Find special products when polynomial products are mixed together.

Key Terms
Use the vocabulary terms listed below to complete each statement in Exercises 1–4.

    binomials   **difference**   **FOIL**   **square**

1. The expression $(x+5)(x^2+2)$ is a product of two ___________________.

2. The multiplication $(x+y)(a+b)=xa+xb+ya+yb$ illustrates the

    ___________________ method.

3. The expression $(x+2)(x-2)$ is the product of the sum and the ___________________

    of the same two terms.

4. The expression $(x+3)^2$ is the ___________________ of a binomial.

Objective a   Multiply two binomials mentally using the FOIL method.

Multiply. Try to write only the answer. If you need more steps, be sure to use them.

5. $(x+2)(x^3+3)$ 5.___________________

6. $(t+6)(t-7)$ 6.___________________

7. $(2a-5)(a+3)$ 7.___________________

8. $(5x-7)(x-1)$

8. _______________

9. $(t-0.2)(t+0.2)$

9. _______________

10. $(-2n+3)(n+8)$

10. _______________

11. $(1+5x)(1-8x)$

11. _______________

12. $(x^2+6)(x^4-2)$

12. _______________

13. $(n^{12}+5)(n^{12}-5)$

13. _______________

14. $(9x^6+2)(x^6+9)$

14. _______________

Objective b   Multiply the sum and the difference of two terms mentally.

Multiply mentally, if possible. If you need extra steps, be sure to use them.

15. $(x+3)(x-3)$

15. _______________

16. $(3x+7)(3x-7)$

16. _______________

17. $(4t^3-5)(4t^3+5)$

17. _______________

18. $(n^8-n^3)(n^8+n^3)$

18. _______________

19. $(x^5+2x)(x^5-2x)$

19. _______________

20. $(3x^6+1.2)(3x^6-1.2)$

20. _______________

110

Objective c   Square a binomial mentally.

Multiply mentally, if possible. If you need extra steps, be sure to use them.

21. $(x+5)^2$

21. ______________

22. $(2x^2+3)^2$

22. ______________

23. $\left(t-\dfrac{1}{3}\right)^2$

23. ______________

24. $(4+x^2)^2$

24. ______________

25. $(6-2x^5)^2$

25. ______________

26. $\left(a-\dfrac{3}{4}\right)^2$

26. ______________

Objective d   Find special products when polynomial products are mixed together.

Multiply mentally, if possible.

27. $(5-3x^2)^2$

27. ______________

28. $-2x^2(3x^2-4x-7)$

28. ______________

29. $\left(6t^2+\dfrac{1}{5}\right)\left(6t^2-\dfrac{1}{5}\right)$

29. ______________

30. $(-5a+8)(5a+8)$

30. ______________

31. $(y+2)(y^2-2y+4)$

31. ______________

32. $(7t^4+10)^2$

32. ______________

33. $4^2 + 5^2$; $(4+5)^2$

33. _________________

34. $10^2 - 2^2$; $(10-2)^2$

34. _________________

Find the total area of all the shaded rectangles.

35. 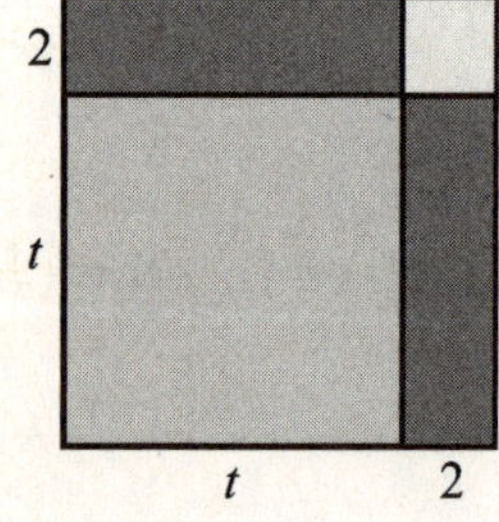

35. _________________

36. 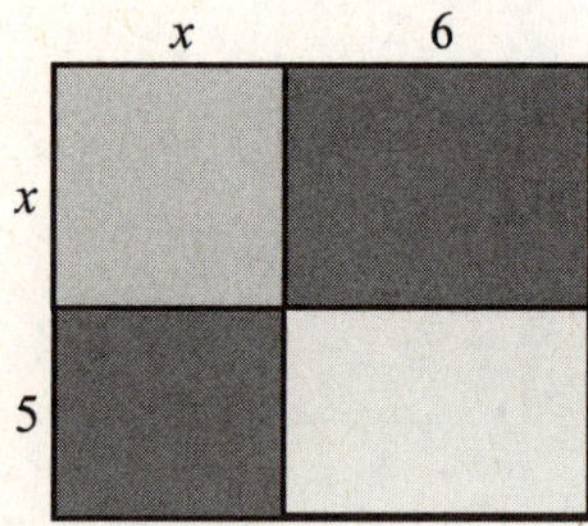

36. _________________

112

Chapter 4 POLYNOMIALS: OPERATIONS

4.7 Operations with Polynomials in Several Variables

Learning Objectives
a   Evaluate a polynomial in several variables for given values of the variables.
b   Identify the coefficients and the degrees of the terms of a polynomial and the degree of a polynomial.
c   Collect like terms of a polynomial.
d   Add polynomials.
e   Subtract polynomials.
f   Multiply polynomials.

Key Terms
Use the vocabulary terms listed below to complete each statement in Exercises 1–2.

    degree   **like terms**

1.   The _________________________ of a term is the sum of the exponents of the variables.

2.   _________________________ have exactly the same variables with exactly the same exponents.

Objective a   Evaluate a polynomial in several variables for given values of the variables.

Evaluate the polynomial when $x = -2,\ y = 3,\ and\ z = -1.$

3. $x^2 + 2xy^2 - y^3$ **3.** ________________

4. $3xyz - 2z^2$ **4.** ________________

113

5. The polynomial equation $C = 370 + 21.6w(1-p)$ can be used to estimate the number of daily calories needed to maintain weight for a person with mass w, in kilograms, and body fat percentage p (written in decimal form). How many calories are needed to maintain the weight of a 65-kg adult with a 25% body fat percentage?

5. _______________

6. The altitude h, in ft, of a launched object is given by the polynomial equation $h = h_0 + vt - 16t^2$, where h_0 is the height, in ft, from which the launch occurs, v is the initial upward speed (or velocity), in ft/sec, and t is the number of seconds for which the object is airborne. A stone is thrown upward with an initial speed of 40 ft/sec from the top of a 300-ft cliff. How high above the base of the cliff will the stone be after 3 sec?

6. _______________

Objective b   Identify the coefficients and the degrees of the terms of a polynomial and the degree of a polynomial.

Identify the coefficient and the degree of each term of the polynomial. Then find the degree of the polynomial.

7. $2xy^2 - 3xy + 4x^4 - 7$

7. _______________

8. $26x^3y^4 - 10x^2y + 15$

8. _______________

114

Objective c   Collect like terms of a polynomial.

Collect like terms.

9. $2c + d - c - 4d$

9. ____________________

10. $8x^3 + 3x^2 y - 5xy^2$

10. ____________________

11. $4xy + 3xz + 18xy + 10xz$

11. ____________________

12. $4cd^2 - 8c^2 d - 2c^2 d - 3cd^2$

12. ____________________

Objective d   Add polynomials.

Add.

13. $\left(-2x^2 + 5xy + y^2\right) + \left(x^2 - 7xy - 6y^2\right)$

13. ____________________

14. $\left(ab + 2a + 3b\right) + \left(a - 5b - 6ab\right) + \left(7b - 4a - ab\right)$

14. ____________________

15. $\left(8u^2 + 3uv + v^2\right) + \left(-10u^2 - 8uv - v^2\right) + \left(u^2 + uv - v^2\right)$

15. ____________________

16. $\left(r^2 s^3 + 5r^3 s^2 - 6rs - 7\right) + \left(2r^3 s^2 - 7r^2 s^3 + 5rs + 10\right)$

16. ____________________

Objective e   Subtract polynomials.

Subtract.

17. $\left(x^3 + 2y^3\right) - \left(3x^3 y - xy^3 - x^3 + 2y^3\right)$

17. ____________________

18. $\left(rs - tu - 7\right) - \left(rs + 3tu - 10\right)$

18. ____________________

19. $\left(6a^5 b^2 + 8a^4 b - 4b + 10\right) - \left(a^5 b^2 + 8a^4 b - 6a - 2b - 1\right)$

19. ____________________

20. $\left(-3x + 7y + z\right) - \left(-5x + 8y - 10z\right)$

20. ____________________

Objective f   Multiply polynomials.

Multiply.

21. $(4x-3y)(2x+y)$

22. $(t^3+2rs)(t^3-2rs)$

23. $(a^2+a^2b+3)(b^2+1)$

24. $(6-x^2y^2)(8+x^2y^2)$

25. $(c+2d)^2$

26. $(m^2-p^3q^3)^2$

27. $10x(x-3y)(x+3y)$

28. $(x^3+3y^2+2)(x^3-2y^2+7)$

29. $(a+b+5)(a+b-5)$

30. $(a^2+2b+1)^2$

21. _______________

22. _______________

23. _______________

24. _______________

25. _______________

26. _______________

27. _______________

28. _______________

29. _______________

30. _______________

116

Chapter 4 POLYNOMIALS: OPERATIONS

4.8 Division of Polynomials

Learning Objectives
a   Divide a polynomial by a monomial.
b   Divide a polynomial by a divisor that is a binomial.

Key Terms
Use the vocabulary terms listed below in Exercises 1–4 to label each part of the division shown.

       dividend   **divisor**   **quotient**   **remainder**

1. _______________

2. _______________

3. _______________

4. _______________

$$
\begin{array}{r}
x + 2 \quad \text{①} \\
x + 3 \,\overline{)\, x^2 + 5x + 10} \quad \text{③} \\
\underline{x^2 + 3x} \\
2x + 10 \\
\underline{2x + 6} \\
4 \quad \text{④}
\end{array}
$$

where ② → $x + 3$

Objective a   Divide a polynomial by a monomial.

Divide and check.

5. $\dfrac{30x^7}{5}$

5. _______________

6. $\dfrac{-18x^5}{3x^2}$

6. _______________

7. $\dfrac{-51c^3 d^2}{-3cd}$

7. _______________

117

8. $\dfrac{48x^4 - 36x^3 - 16x^2 + 60}{4}$

8. _______________

9. $\dfrac{13x^5 - 16x^4 + 20x^2}{x}$

9. _______________

10. $\left(60t^6 + 36t^4 - 18t^2\right) \div \left(6t\right)$

10. _______________

11. $\left(22a^7 - 26a^5 + 32a^3\right) \div \left(-2a\right)$

11. _______________

12. $\left(100x^6 - 80x^5 - 45x^4\right) \div \left(-10x^4\right)$

12. _______________

13. $\dfrac{12a^2 - 6a + 1}{3}$

13. _______________

14. $\dfrac{8a^2b^3 + 10ab^2 - 4ab}{2ab}$

14. _______________

118

Objective b   Divide a polynomial by a divisor that is a binomial.

Divide.

15. $\left(x^2 + 10x + 25\right) \div \left(x + 5\right)$

15. _________________

16. $\left(x^2 + 6x - 9\right) \div \left(x - 3\right)$

16. _________________

17. $\left(x^2 + 4x + 3\right) \div \left(x - 1\right)$

17. _________________

18. $\dfrac{x^2 - 100}{x - 10}$

18. _________________

19. $\dfrac{x^4+1}{x+1}$

19. _______________

20. $\dfrac{3x^3-7x^2-21x-10}{3x+2}$

20. _______________

21. $\left(x^6-x^3-30\right)\div\left(x^3-6\right)$

21. _______________

22. $\left(20x^3+x^2-9x+6\right)\div\left(4x+1\right)$

22. _______________

23. $\left(4x^4-5x^2-8x+1\right)\div\left(2x-3\right)$

23. _______________

24. $\left(x^3+x^2+1\right)\div\left(x-1\right)$

24. _______________

120

Chapter 5 POLYNOMIALS: FACTORING

5.1 Introduction to Factoring

Learning Objectives
a   Find the greatest common factor, the GCF, of monomials.
b   Factor polynomials when the terms have a common factor, factoring out the greatest common factor.
c   Factor certain expressions with four terms using factoring by grouping.

Key Terms
Use the vocabulary terms listed below to complete each statement in Exercises 1–4. Terms may be used more than once.

    factor   **factoring by grouping**   **factorization**

1. To _________________ a polynomial is to express it as a product.

2. A(n) _________________ of a polynomial P is a polynomial that can be used to express P as a product.

3. A(n) _________________ of a polynomial is an expression that names that polynomial as a product.

4. Certain polynomials with four terms can be factored using _________________.

Objective a   Find the greatest common factor, the GCF, of monomials.

Find the GCF.

5. $x^3,\ -10x$ 5. _______________

6. $4x^5,\ x^3$ 6. _______________

7. $3x,\ 3x^5,\ 12$ 7. _______________

8. $-11x^4y^3$, $33x^3y^5$, $55x^2y$

8. ______________

9. $-15x$, $13x^2$, $18x^7$

9. ______________

10. p^3q^3, p^7q^2, p^4q^6, p^5q^8

10. ______________

Objective b Factor polynomials when the terms have a common factor, factoring out the greatest common factor.

Factor. Check by multiplying.

11. $x^2 - 10x$

11. ______________

12. $3x^6 + 12x^4$

12. ______________

13. $10x^5 - 5x^2$

13. ______________

14. $4x^2 + 4x - 20$

14. ______________

15. $11x^4y^3 - 33x^3y^5 - 55x^2y$

15. ______________

122

16. $8x^5 + 10x^4 - 7x^3$ 16. ______________

17. $p^3 q^4 + p^3 q^2 - p^4 q^3 - p^6 q^8$ 17. ______________

18. $3x^8 - 30x^6 + 15x^5 + 18x^3$ 18. ______________

19. $1.4x^4 - 3.5x^3 + 4.2x^2 + 7.0x$ 19. ______________

20. $\dfrac{3}{2}x^7 + \dfrac{7}{2}x^5 - \dfrac{1}{2}x^3 + \dfrac{1}{2}x^2$ 20. ______________

Objective c   Factor certain expressions with four terms using factoring by grouping.

Factor.

21. $x^3(x+1) + 3(x+1)$ 21. ______________

22. $7p^2(4p-5) - (4p-5)$ 22. ______________

Factor by grouping.

23. $x^3 + 2x^2 + 5x + 10$ 　　　　　　　　**23.** _______________

24. $2z^3 + 2z^2 + 3z + 3$ 　　　　　　　　**24.** _______________

25. $8p^3 - 20p^2 + 10p - 25$ 　　　　　　　**25.** _______________

26. $20x^3 - 15x^2 + 8x - 6$ 　　　　　　　**26.** _______________

27. $3x^3 - 6x^2 - x + 2$ 　　　　　　　　**27.** _______________

28. $x^3 + 3x^2 - 7x - 21$ 　　　　　　　　**28.** _______________

Chapter 5 POLYNOMIALS: FACTORING

5.2 Factoring Trinomials of the Type $x^2 + bx + c$

> **Learning Objective**
> a   Factor trinomials of the type $x^2 + bx + c$ by examining the constant term c.

Key Terms
Use the vocabulary terms listed below to complete each statement in Exercises 1–4.

leading coefficient	negative	positive	prime

1. In the expression $ax^2 + bx + c,$ a is called the ___________________.

2. A(n) ___________________ polynomial cannot be factored further.

3. When the constant term of a trinomial is ___________________, the constant terms of the binomial factors have the same sign.

4. When the constant term of a trinomial is ___________________, the constant terms of the binomial factors have opposite signs.

Objective a   Factor trinomials of the type $x^2 + bx + c$ by examining the constant term c.

Factor. Remember that you can check by multiplying.

5. $x^2 + 9x + 18$ 5. ___________________

Pairs of Factors	Sums of Factors

125

6. $x^2 - 8x + 16$

Pairs of Factors	Sums of Factors

6. _______________

7. $a^2 + a - 20$

Pairs of Factors	Sums of Factors

7. _______________

8. $y^2 - 9y + 8$

Pairs of Factors	Sums of Factors

8. _______________

9. $t^2 - 13t + 30$

9. _______________

10. $p^2 - 5p - 14$

10. _______________

11. $x^2 + 3x + 1$

11. _______________

12. $x^3 - 5x^2 + 6x$ **12.** ___________________

13. $x^3 - 3x^2 - 40x$ **13.** ___________________

14. $16x + 60 + x^2$ **14.** ___________________

15. $t^4 + 7t^2 + 10$ **15.** ___________________

16. $15 - 5p + p^2$ **16.** ___________________

17. $x^2 + 22x + 121$ **17.** ___________________

18. $27 - 6a - a^2$ **18.** ___________________

19. $x^4 - 13x^3 - 30x^2$ **19.** ___________________

127

20. $y^2 - 16y + 60$

20. _______________

21. $75 - 28t + t^2$

21. _______________

22. $x^2 - 0.1x - 0.06$

22. _______________

23. $60 - 7t - t^2$

23. _______________

24. $p^2 - pq - 20q^2$

24. _______________

25. $m^2 - 5mn - 66n^2$

25. _______________

26. $5a^8 - 15a^7 - 140a^6$

26. _______________

128

Chapter 5 POLYNOMIALS: FACTORING

5.3 Factoring $ax^2 + bx + c$, $a \neq 1$: The FOIL Method

Learning Objective
a   Factor trinomials of the type $ax^2 + bx + c$, $a \neq 1$, using the FOIL method.

Key Terms
Use the vocabulary terms listed below to complete each statement in Exercises 1–4.

First	Inside	Last	Outside

1. In the product $(x-3)(2x+1)$, x and $2x$ are the ___________________ terms.

2. In the product $(x-3)(2x+1)$, -3 and 1 are the ___________________ terms.

3. In the product $(x-3)(2x+1)$, -3 and $2x$ are the ___________________ terms.

4. In the product $(x-3)(2x+1)$, x and 1 are the ___________________ terms.

Objective a   Factor trinomials of the type $ax^2 + bx + c$, $a \neq 1$, using the FOIL method.

Factor.

5. $2x^2 - 5x - 12$ 5. ___________________

6. $5x^2 - 32x + 12$ 6. ___________________

7. $6x^2 - 17x + 5$ 7. ___________________

129

8. $3x^2 - 2x - 1$

8. _______________

9. $16x^2 + 8x - 15$

9. _______________

10. $18x^2 + 15x + 2$

10. _______________

11. $21 - 58x + 21x^2$

11. _______________

12. $25x^2 - 40x + 16$

12. _______________

13. $7 - 5x - 2x^2$

13. _______________

14. $8x^2 - 28x - 36$

14. _______________

130

15. $15x^2 + 55x + 50$ 15. _______________

16. $-1 + 3x^2 - 2x$ 16. _______________

17. $16x^2 - 72x - 40$ 17. _______________

18. $21t - 6t^2 - 9$ 18. _______________

19. $10p^4 - 21p^3 - 10p^2$ 19. _______________

20. $14a^4 + 23a^2 + 3$ 20. _______________

21. $81y^2 + 36y + 4$

21. _______________

22. $16x^2 + 21x + 25$

22. _______________

23. $6m^2 - 11mn - 10n^2$

23. _______________

24. $20a^2 + 29ab + 6b^2$

24. _______________

25. $14p^2 + 23pq + 3q^2$

25. _______________

26. $16x^2 - 8xy - 8y^2$

26. _______________

132

Chapter 5 POLYNOMIALS: FACTORING

5.4 Factoring $ax^2 + bx + c$, $a \neq 1$: The ac-Method

Learning Objective
a   Factor trinomials of the type $ax^2 + bx + c$, $a \neq 1$, using the ac-method.

Key Terms
Use the vocabulary terms listed below to complete the steps for factoring using the ac-method in Exercises 1–6.

common factor	**grouping**	**leading coefficient**
multiplying	**split**	**sum**

1. Factor out a(n) _____________________, if any.

2. Multiply the _____________________ a and the constant c.

3. Try to factor the product ac so that the _____________________ of the factors is b.

4. _____________________ the middle term.

5. Factor by _____________________.

6. Check by _____________________.

Objective a   Factor trinomials of the type $ax^2 + bx + c$, $a \neq 1$, using the ac-method.

Factor. Note that the middle term has already been split.

7. $x^2 + 4x + 3x + 12$ 7. _____________________

8. $6x^2 + 4x + 15x + 10$ 8. _____________________

133

9. $24y^2 - 20y + 6y - 5$

9. _________________

10. $2x^4 - 4x^2 - 7x^2 + 14$

10. _________________

Factor using the ac-method.

11. $3x^2 + 13x - 10$

11. _________________

12. $8x^2 + 26x + 15$

12. _________________

13. $4x^2 + 8x - 21$

13. _________________

14. $4x^2 - 16x - 9$

14. _________________

15. $14x^2 - 65x + 9$

15. _________________

16. $25a^2 - 30a + 9$

16. _________________

134

17. $3 - 5a - 2a^2$

17. _______________

18. $15x^2 + 99x - 42$

18. _______________

19. $15t^2 - 22t - 9$

19. _______________

20. $4x^2 - 7x - 2$

20. _______________

21. $6t^3 + 19t^2 + 15t$

21. _______________

22. $26y - 6y^2 - 8$

22. _______________

23. $12p^4 - 4p^3 - 5p^2$

23. _______________

24. $105x^3 - 50x^2 + 5x$

24. _______________

25. $8x^4 - 26x^2 + 15$

25. _________________

26. $16x^2 + 72x + 81$

26. _________________

27. $15y^2 + 10y + 18$

27. _________________

28. $15a^2 + 11ab + 2b^2$

28. _________________

29. $6s^2 + 20st + 14t^2$

29. _________________

30. $8x^2 - 28xy + 24y^2$

30. _________________

136

Chapter 5 POLYNOMIALS: FACTORING

5.5 Factoring Trinomial Squares and Differences of Squares

Learning Objectives
a   Recognize trinomial squares.
b   Factor trinomial squares.
c   Recognize differences of squares.
d   Factor differences of squares, being careful to factor completely.

Key Terms

Use the vocabulary terms listed below to complete each statement in Exercises 1–4.

    difference of squares   **factored completely**

    sum of squares   **trinomial square**

1.   The expression $x^2 + 10x + 25$ is a(n) ___________________.

2.   The expression $9x^2 - 16$ is a(n) ___________________.

3.   The expression $y^2 + 81$ is a(n) ___________________.

4.   When no factor can be factored further, we have ___________________.

Objective a   Recognize trinomial squares.

Determine whether each of the following is a trinomial square.

5. $x^2 + 10x - 25$ 5. ___________________

6. $4x^2 - 12x + 9$ 6. ___________________

137

Objective b   Factor trinomial squares.

Factor completely. Remember to look first for a common factor and to check by multiplying.

7. $x^2 - 6x + 9$

7. ______________

8. $36 + 12x + x^2$

8. ______________

9. $m^4 + 10m^2 + 25$

9. ______________

10. $12a^2 - 84a + 147$

10. ______________

11. $x^3 - 20x^2 + 100x$

11. ______________

12. $64 - 80x + 25x^2$

12. ______________

13. $4m^2 + 20mn + 25n^2$

13. ______________

14. $4p^4 + 16p^2 + 16$

14. ______________

15. $x^2 - 8xy + 16y^2$

15. ______________

16. $81a^2 - 18ab + b^2$

16. ______________

138

Objective c   Recognize differences of squares.

Determine whether each of the following is a difference of squares.

17. $x^2 - 100$

17. _______________

18. $-4x^2 + 10y^2$

18. _______________

19. $x^2 + 16$

17. _______________

Objective d   Factor differences of squares, being careful to factor completely.

Factor completely. Remember to look first for a common factor.

20. $a^2 - 81$

20. _______________

21. $-100 + y^2$

21. _______________

22. $64x^2 - y^2$

22. _______________

23. $4m^2 - 49n^2$

23. ______________

24. $18x^2 - 32$

24. ______________

25. $0.04a^2 - 0.0025$

25. ______________

26. $36p^4 - 1$

26. ______________

27. $3x^4 - 48$

27. ______________

28. $x^{16} - 81$

28. ______________

29. ______________

29. $9 - \dfrac{1}{16}x^2$

140

Chapter 5 POLYNOMIALS: FACTORING

5.6 Factoring Sums or Differences of Cubes

Learning Objectives
a   Factor sums or differences of cubes.

Key Terms
In Exercises 1–3, match the term with the appropriate example in the column on the right.

1. _____ difference of cubes   **a)** $t^2 - s^2$

2. _____ difference of squares   **b)** $5^3 - 8^3$

3. _____ sum of cubes   **c)** $z^3 + 10^3$

Objective a   Factor sums and differences of cubes.

Factor.

4. $x^3 + 125$ 4. _________________

5. $64t^3 - 1$ 5. _________________

6. $8 + 27y^3$ 6. _________________

7. $z^3 + w^3$ 7. _________________

8. $6 - 6p^3$ 8. _________________

9. $d^3 - 0.008$ 9. _________________

10. $3a^3 + 24b^3$

10. _______________

11. $x^3 - \dfrac{1}{27}$

11. _______________

12. $pq^3 - p^4$

12. _______________

13. $1250t^6 + 10s^6$

13. _______________

14. $3m^5 + 3000m^2$

14. _______________

15. $x^6 - 1$

15. _______________

16. $y^6 - 64z^6$

16. _______________

17. $x^{15} + x^6y^{12}$

17. _______________

18. $8b^3 + 27$

18. _______________

19. $r^6 - 64$

19. _______________

20. $a^{12} + b^{15}c^6$

20. _______________

142

Chapter 5 POLYNOMIALS: FACTORING

5.7 Factoring: A General Strategy

Learning Objective
a   Factor polynomials completely using any of the methods considered in this chapter.

Key Terms
Use the vocabulary terms listed below to complete each statement in Exercises 1–4.

    completely   **difference**   **grouping**   **square**

1.   When factoring a polynomial with two terms, determine whether you have

     a(n) _________________ of squares.

2.   When factoring a polynomial with three terms, determine whether the trinomial is

     a(n) _______________.

3.   When factoring a polynomial with four terms, try factoring by _________________.

4.   Always factor _________________.

Objective a   Factor polynomials completely using any of the methods considered in this chapter.

Factor completely.

5. $5t^2 - 20$ 5. _______________

6. $x^2 - 18x + 81$ 6. _______________

7. $6x^2 - 11x - 10$ 7. _______________

8. $x^3 - 3x^2 - x + 3$

8. _______________

9. $x^4 + 5x^2 - 4x^3 - 20x$

9. _______________

10. $6x^3 + 4x^2 - 10x$

10. _______________

11. $n^2 + 1$

11. _______________

12. $24a^2 - 6$

12. _______________

13. $2x^4 + 12x^3 + 18x^2$

13. _______________

14. $x^2 + 3x + 7$

14. _______________

15. $25 - 45x - 10x^2$

15. _______________

16. $3x^5 - 3x$

16. _______________

144

17. $3p^3q + 12pq^3$

17. ______________

18. $4x\left(u^2 + 3v\right) - \left(u^2 + 3v\right)$

18. ______________

19. $a^2 - a + ab - b$

19. ______________

20. $20m^3n^5 + 5m^2n^3 - 15m^4n^4$

20. ______________

21. $9c^2 + 4d^2 - 12cd$

21. ______________

22. $25x^2z^2 + 40xyz + 16y^2$

22. ______________

23. $m^2 - 4mn - 60n^2$

23. ______________

24. $a^2b^2 + 2ab - 15$

24. ______________

25. $x^5y^2 + 4x^4y - 32x^3$

25. ______________

145

26. $r^2 - \dfrac{1}{4}t^2$

26. _______________

27. $c^8 d^4 - 81$

27. _______________

28. $49m^4 - 70m^3 n + 25m^2 n^2$

28. _______________

146

Chapter 5 POLYNOMIALS: FACTORING

5.8 Solving Quadratic Equations by Factoring

Learning Objectives
a   Solve equations (already factored) using the principle of zero products.
b   Solve equations by factoring and then using the principle of zero products.

Key Terms
Use the vocabulary terms listed below to complete each statement in Exercises 1–4.

> **factor**   **products**   **quadratic equation**   **zero**

1. $3x^2 - 5x + 3 = 0$ is an example of a(n) _____________________.

2. The principle of zero _____________________ states that a product is 0 if and only if one or both of the factors is 0.

3. To use the principle of zero products, you must have _____________________ on one side of the equation.

4. To use the principle of zero products, we _____________________ a quadratic polynomial.

Objective a   Solve equations (already factored) using the principle of zero products.

Solve using the principle of zero products.

5. $(x+3)(x-10) = 0$ 5. _____________________

6. $(x+5)(x+7) = 0$ 6. _____________________

147

7. $x(x+15)=0$

7. _______________

8. $0=x(x-8)$

8. _______________

9. $(4x+3)(2x-10)=0$

9. _______________

10. $(6x-11)(8x-3)=0$

10. _______________

11. $\left(\dfrac{1}{2}-3x\right)\left(\dfrac{1}{3}-4x\right)=0$

11. _______________

12. $(0.1x+0.2)(0.5x-15)=0$

12. _______________

148

Objective b   Solve equations by factoring and then using the principle of zero products.

Solve by factoring and using the principle of zero products. Remember to check.

13. $x^2 + 5x + 4 = 0$ **13.** _______________

14. $x^2 + 4x - 21 = 0$ **14.** _______________

15. $x^2 - 10x = 0$ **15.** _______________

16. $x^2 + 12x = 0$ **16.** _______________

17. $x^2 = 25$ **17.** _______________

18. $16x^2 - 1 = 0$ **18.** _______________

149

19. $0 = 4x + x^2 + 4$

19. _______________

20. $3x^2 = 10x$

20. _______________

21. $3x^2 + 5x = 2$

21. _______________

22. $20n^2 = 11n + 3$

22. _______________

23. $2y^2 + 16y + 30 = 0$

23. _______________

24. $t(3t + 5) = 28$

24. _______________

150

Find the x-intercepts for the graph of the equation. (The grids are intentionally not included.)

25.

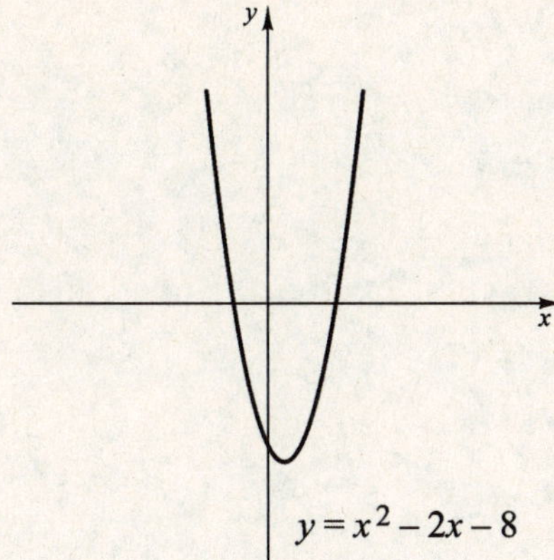

25. _______________

26.

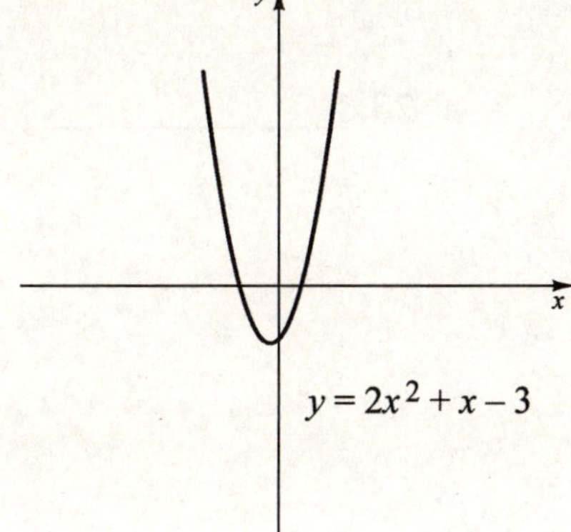

26. _______________

151

27. Use the following graph to solve $x^2 - x - 2 = 0$.

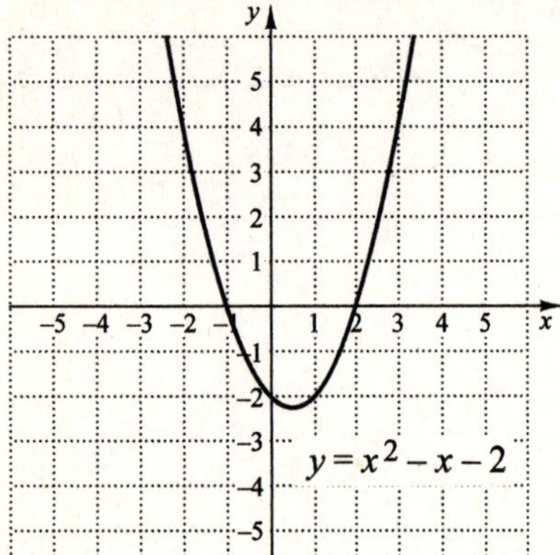

28. Use the following graph to solve $-x^2 - 2x + 3 = 0$.

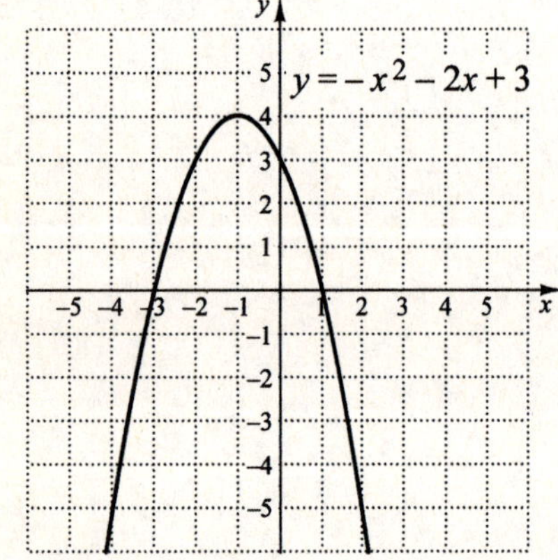

152

Chapter 5 POLYNOMIALS: FACTORING

5.9 Applications of Quadratic Equations

Learning Objectives
a   Solve applied problems involving quadratic equations that can be solved by factoring.

Key Terms
Use the vocabulary terms listed below to complete each statement in Exercises 1–4.

hypotenuse	legs	Pythagorean	right

1. A triangle that has a $90°$ angle is a(n) ___________________ triangle.

2. In a right triangle, the side opposite the $90°$ angle is the ___________________.

3. The sides that form the $90°$ angle in a right triangle are called ___________________.

4. The ___________________ theorem states that, in a right triangle, $a^2 + b^2 = c^2$.

Objective a   Solve applied problems involving quadratic equations that can be solved by factoring.

Solve.

5. The length of a rectangular table is 5 ft greater than the width. The area of the table is 24 ft². Find the length and the width.

5. ___________________

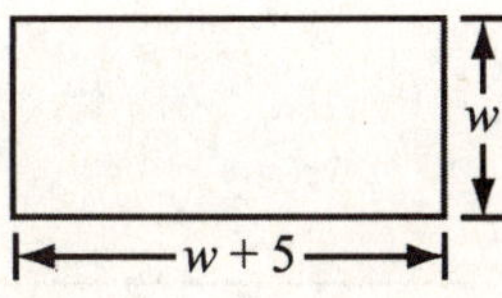

6. A rectangular serving tray is twice as long as it is wide. The area of the tray is 338 in². Find the dimensions of the

6. ___________________

153

tray.

7. A triangle is 8 cm wider than it is tall. The area is 42 cm².   7. _______________
   Find the height and the base.

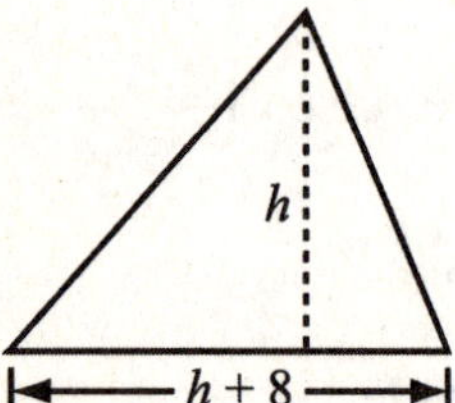

8. A triangular garden has a base twice as long as its height.   8. _______________
   The garden has an area of 49 m². Find the base and the
   height.

9. A soccer league has 12 teams. What is the total number of   9. _______________
   games to be played if each team plays every other team
   twice? Use $x^2 - x = N$, where N is the number of games
   and x is the number of teams.

10. A softball league plays a total of 210 games. How many   10. _______________
    teams are in the league if each team plays every other team
    twice? Use $x^2 - x = N$, where N is the number of games
    and x is the number of teams.

154

11. There are 30 people at a party. How many handshakes are possible? Use $N = \dfrac{1}{2}\left(x^2 - x\right)$, where N is the number of handshakes and x is the number of people.

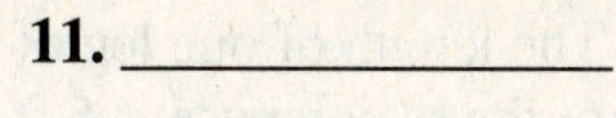

12. Everyone at a meeting shook hands with each other. There were 276 handshakes in all. How many people were at the meeting? Use $N = \dfrac{1}{2}\left(x^2 - x\right)$, where N is the number of handshakes and x is the number of people.

13. During a toast at a party, there were 66 "clinks" of glasses. How many people took part in the toast? Use $N = \dfrac{1}{2}\left(x^2 - x\right)$, where N is the number of clinks and x is the number of people.

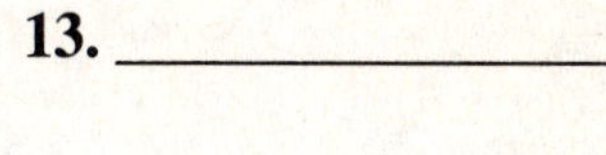

14. The product of the page numbers on two facing pages of a book is 110. Find the page numbers.

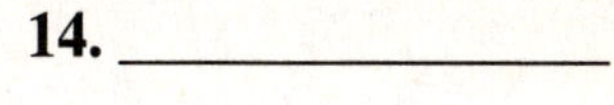

15. The product of two consecutive even integers is 288. Find the integers.

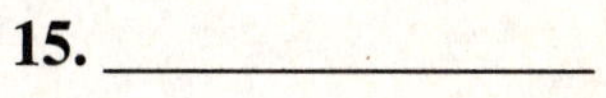

16. The product of two consecutive odd integers is 195. Find the integers.

17. The length of one leg of a right triangle is 15 ft. The length
of the hypotenuse is 5 ft longer than the other leg. Find the
length of the hypotenuse and the other leg.

17. ________________

18. A metal sign is a right triangle that is 30 in. tall and has a
hypotenuse of length 34 in. If the metal costs \$0.10 per
square inch, find the cost of the sign.

18. ________________

19. Johann landed his Cessna 172, traveling 13,000 ft over a
12,000-ft horizontal distance. From what altitude did the
descent begin?

19. ________________

20. The guy wire on a tower is 10 ft longer than the height of
the tower. If the guy wire is anchored 50 ft from the foot
of the tower, how tall is the tower?

20. ________________

21. A model rocket is launched with an initial velocity of 160
ft/sec. Its height h, in feet, after t seconds is given by the
formula $h = 160t - 16t^2$. After how many seconds will the
rocket first reach a height of 336 ft?

21. ________________

22. The sum of the squares of two consecutive odd positive
integers is 202. Find the integers

22. ________________

156

Chapter 6 RATIONAL EXPRESSIONS AND EQUATIONS

6.1 Multiplying and Simplifying Rational Expressions

> **Learning Objectives**
> a   Find all numbers for which a rational expression is not defined.
> b   Multiply a rational expression by 1, using an expression such as A/A.
> c   Simplify rational expressions by factoring the numerator and the denominator and removing factors of 1.
> d   Multiply rational expressions and simplify.

Key Terms
Use the vocabulary terms listed below to complete each statement in Exercises 1–4.

      equivalent   **multiply**   **rational**   **simplify**

1.   A quotient, or ratio, of polynomials is a(n) ___________________ expression.

2.   Expressions that have the same value for all allowable replacements are called

    ___________________ expressions.

3.   To ___________________ rational expressions, multiply numerators and multiply

    denominators.

4.   To ___________________ rational expressions, factor the numerator and the

    denominator and "remove" a factor of 1.

Objective a   Find all numbers for which a rational expression is not defined.

Find all numbers for which a rational expression is not defined.

5. $\dfrac{-1}{5x}$ 5. ___________________

6. $\dfrac{3}{x+7}$ 6. ___________________

7. $\dfrac{2}{4a-5}$

7.____________________

8. $\dfrac{5x+1}{x^2-x-6}$

8.____________________

9. $\dfrac{n^2-n}{n^2-16}$

9.____________________

10. $\dfrac{p+7}{20}$

10.____________________

Objective b   Multiply a rational expression by 1, using an expression such as A/A.

Multiply. Do not simplify. Note that in each case you are multiplying by 1.

11. $\dfrac{5a^2}{5a^2}\cdot\dfrac{6c^2}{7d^4}$

11.____________________

12. $\dfrac{y-6}{y+1}\cdot\dfrac{y-7}{y-7}$

12.____________________

158

Name: Date:
Instructor: Section:

Objective c   Simplify rational expressions by factoring the numerator and the denominator and removing factors of 1.

Simplify.

13. $\dfrac{4x^5}{20x}$

13. _______________

14. $\dfrac{30a^8b^5}{12ab^4}$

14. _______________

15. $\dfrac{8m-40}{8m}$

15. _______________

16. $\dfrac{4x^2+8x}{12x^3+4x}$

16. _______________

17. $\dfrac{t^2-1}{t^2-10t+9}$

17. _______________

18. $\dfrac{a^2+a-2}{a^2+2a-3}$

18. _______________

19. $\dfrac{x^2+25}{x+5}$

19. _______________

20. $\dfrac{9x^2-36}{3x^2-12}$

20. _______________

21. $\dfrac{2x^2+14x+20}{6x^2+12x-90}$

21. _______________

22. $\dfrac{c^2-12c+36}{c^2-36}$

22. _______________

23. $\dfrac{a-6}{6-a}$

23. _______________

24. $\dfrac{x^2-100}{10-x}$

24. _______________

160

Name: Date:
Instructor: Section:

Objective d   Multiply rational expressions and simplify.

Multiply and simplify.

25. $\dfrac{5x^6}{3x}\cdot\dfrac{9}{2x}$

25. _________________

26. $\dfrac{4a}{b^3}\cdot\dfrac{3b}{10a^2}$

26. _________________

27. $\dfrac{x^2+6x+5}{x^2-4x+3}\cdot\dfrac{x-3}{x+5}$

27. _________________

28. $\dfrac{t^2-25}{3t^3}\cdot\dfrac{t^2-2t}{t^2+7t+10}$

28. _________________

161

29. $\dfrac{6x^3}{5x^2+30x+45}\cdot\dfrac{5x+15}{3x}$

29. __________________

30. $\dfrac{x^4-16}{x^4-81}\cdot\dfrac{x^2+9}{x^2+4}$

30. __________________

31. $\dfrac{(m-7)^3}{(m+5)^3}\cdot\dfrac{m^2+10m+25}{m^2-14m+49}$

31. __________________

32. $\dfrac{3y^2-27}{2y^2-128}\cdot\dfrac{8y+64}{6y-6}$

32. __________________

162

Chapter 6 RATIONAL EXPRESSIONS AND EQUATIONS

6.2 Division and Reciprocals

Learning Objectives
a   Find the reciprocal of a rational expression.
b   Divide rational expressions and simplify.

Key Terms
Use the vocabulary terms listed below to complete each statement in Exercises 1–2.

interchange   **multiply**

1.   To find the reciprocal of a rational expression, _____________________ the numerator
and the denominator.

2.   To divide by a rational expression, _____________________ by its reciprocal.

Objective a   Find the reciprocal of a rational expression.

Find the reciprocal.

3. $\dfrac{3}{y}$ 3. _____________________

4. $x^2 - 9$ 4. _____________________

5. $\dfrac{1}{n+3}$ 5. _____________________

6. $\dfrac{t^2 + 3t + 7}{t^2 - t - 5}$ 6. _____________________

Objective b   Divide rational expressions and simplify.

Divide and simplify.

7. $\dfrac{3}{2} \div \dfrac{6}{5}$

7. _______________

8. $\dfrac{5}{t} \div \dfrac{15}{t}$

8. _______________

9. $\dfrac{x^2}{y^3} \div \dfrac{x^5}{y}$

9. _______________

10. $\dfrac{n+3}{n-5} \div \dfrac{n-3}{n+1}$

10. _______________

11. $\dfrac{t^2-4}{t^2} \div \dfrac{t+2}{t-2}$

11. _______________

12. $\dfrac{y-7}{18} \div \dfrac{y-7}{3}$

12. _______________

164

13. $\dfrac{4x+16}{21} \div \dfrac{x+4}{14}$

13. _______________

14. $\dfrac{-8+6x}{15} \div \dfrac{9x-12}{5}$

14. _______________

15. $\dfrac{5a+5}{a-3} \div \dfrac{a+1}{a-7}$

15. _______________

16. $\dfrac{x^2-1}{x^2+1} \div \dfrac{x+1}{x-1}$

16. _______________

17. $\dfrac{c^2-d^2}{8c+8d} \div \dfrac{c+d}{10c}$

17. _______________

18. $\dfrac{r^2+9r}{r^2+7r+12}\div\dfrac{2r}{r+4}$

18. ___________________

19. $\dfrac{4x^2-8x-5}{2x^2-11x-21}\div\dfrac{6x^2-13x-5}{4x^2+8x+3}$

19. ___________________

20. $\dfrac{a^2+3a+2}{a^2+5a+6}\div\dfrac{a^2-2a-3}{a^2+2a-3}$

20. ___________________

21. $\dfrac{t^2-4}{6t-12}\div\dfrac{3t+9}{2t^2+6t+4}$

21. ___________________

22. $\dfrac{x^2-x-12}{2x^2+10x}\div\dfrac{x^2+8x+15}{x^2-x}$

22. ___________________

166

Name: Date:
Instructor: Section:

Chapter 6 RATIONAL EXPRESSIONS AND EQUATIONS

6.3 Least Common Multiples and Denominators

> **Learning Objectives**
> a   Find the LCM of several numbers by factoring.
> b   Add fractions, first finding the LCD.
> c   Find the LCM of algebraic expressions by factoring.

Key Terms

Use the vocabulary terms listed below to complete each statement in Exercises 1–2.

least common denominator   **least common multiple**

1. The expression $12x^2y^3$ is the _______________ of $6xy^2$, $4x^2y$, and $2y^3$.

2. The expression $12x^2y^3$ is the _______________ of $\dfrac{5}{6xy^2}$, $\dfrac{1}{4x^2y}$, and $\dfrac{3x}{2y^3}$.

Objective a   Find the LCM of several numbers by factoring.

Find the LCM.

3. 24, 30   3. _______________

4. 8, 15   4. _______________

5. 6, 15, 20   5. _______________

6. 10, 50, 120   6. _______________

Objective b Add fractions, first finding the LCD.

Add, first finding the LCD. Simplify if possible.

7. $\dfrac{5}{12}+\dfrac{2}{15}$

7. _________________

8. $\dfrac{1}{8}+\dfrac{3}{20}$

8. _________________

9. $\dfrac{5}{18}+\dfrac{11}{24}$

9. _________________

10. $\dfrac{3}{15}+\dfrac{7}{20}+\dfrac{2}{25}$

10. _________________

Objective c Find the LCM of algebraic expressions by factoring.

Find the LCM.

11. $10x^3,\ 30x^5$

11. _________________

168

12. $3y^3,\ 12x^2y,\ 15xy^4$

12. _______________

13. $5(t+7),\ 45(t+7)$

13. _______________

14. $x,\ x+10,\ x-10$

14. _______________

15. $a^2-9,\ a^2-2a-3$

15. _______________

16. $m^3+6m^2+9m,\ m^2-3m$

16. _______________

17. $x+5,\ (x-5)^2,\ x^2-25$

17. _______________

18. $x^2 - x - 30$, $x^2 - 7x + 6$

18. _______________

19. $4 + 5x$, $16 - 25x^2$, $4 - 5x$

19. _______________

20. $8a^2 - 16a$, $2a^2 + 6a - 20$

20. _______________

21. $4y^5 - 24y^4 + 36y^3$, $6y^3 + 12y^2 - 90y$

21. _______________

22. $10x^2 - 10$, $8x + 8$, $6x^3 - 12x^2 + 6x$

22. _______________

170

Chapter 6 RATIONAL EXPRESSIONS AND EQUATIONS

6.4 Adding Rational Expressions

Learning Objective
a   Add rational expressions.

Key Terms
Use the vocabulary terms listed below to complete the steps for adding rational expressions with different denominators in Exercises 1–4.

    equivalent expression   **least common multiple**

    numerators   **simplify**

1.   Find the _____________________ of the denominators.

2.   For each rational expression, find an _____________________ with the LCD.

3.   Add the _____________________.

4.   _____________________ if possible.

Objective a   Add rational expressions.

Add. Simplify if possible.

5. $\dfrac{1}{10} + \dfrac{3}{10}$

5. _______________

6. $\dfrac{4x}{x+5} + \dfrac{2x-3}{x+5}$

6. _______________

171

7. $\dfrac{4}{a}+\dfrac{7}{a^2}$

7._______________

8. $\dfrac{6}{25n}+\dfrac{2}{15n}$

8._______________

9. $\dfrac{8}{c^2 d}+\dfrac{5}{cd^2}$

9._______________

10. $\dfrac{2x+y}{x^2 y}+\dfrac{3x-y}{xy^2}$

10._______________

11. $\dfrac{9}{7t}+\dfrac{5}{t+3}$

11._______________

12. $\dfrac{3a}{a^2-36}+\dfrac{a}{a+6}$

12._______________

172

13. $\dfrac{5}{y-2}+\dfrac{3}{(y-2)^2}$

13. _______________

14. $\dfrac{t+5}{t}+\dfrac{t}{t+5}$

14. _______________

15. $\dfrac{2}{x^2-6x-7}+\dfrac{5}{x^2-2x-3}$

15. _______________

16. $\dfrac{x+1}{x-4}+\dfrac{x-4}{x+1}$

16. _______________

17. $\dfrac{3x}{x^2-4}+\dfrac{5x}{x^2+2x}$

17. _______________

18. $\dfrac{8}{y}+\dfrac{10}{-y}$

18. _______________

19. $\dfrac{3x+2}{x-10}+\dfrac{6x}{10-x}$

19. _______________

20. $\dfrac{n^2}{n-1}+\dfrac{1}{1-n}$

20. _______________

21. $\dfrac{x+1}{x-3}+\dfrac{2x-3}{3-x}+\dfrac{4(2x+1)}{x-3}$

21. _______________

22. $\dfrac{t+2}{(t-5)(t-3)}+\dfrac{3(t-1)}{(t-5)(3-t)}+\dfrac{(t-2)(t-1)}{(5-t)(t-3)}$

22. _______________

23. $\dfrac{3-a}{49-a^2}+\dfrac{a+2}{a-7}$

23. _______________

24. $\dfrac{4}{x^2-5x+4}+\dfrac{5}{x^2-16}$

24. _______________

174

Chapter 6 RATIONAL EXPRESSIONS AND EQUATIONS

6.5 Subtracting Rational Expressions

Learning Objectives
a   Subtract rational expressions.
b   Simplify combined additions and subtractions of rational expressions.

Key Terms
Use the vocabulary terms listed below to complete each statement in Exercises 1–4.

    denominator   **least common denominator**

    multiply   **numerators**

1.   To subtract rational expressions, they must be written with a common

    ___________________.

2.   The least common multiple of the denominators is the ___________________.

3.   To subtract rational expressions when the denominators are the same, subtract the

    ___________________.

4.   When one denominator is the opposite of the other, we ___________________ one

expression by $-1/-1$ to obtain a common denominator.

Objective a   Subtract rational expressions.

Subtract. Simplify if possible.

5. $\dfrac{10}{x} - \dfrac{4}{x}$

5.   ________________

6. $\dfrac{n}{n-10} - \dfrac{10}{n-10}$

6.   ________________

7. $\dfrac{2x+5}{x^2+6x-7}-\dfrac{x-2}{x^2+6x-7}$

7. _________________

8. $\dfrac{t-3}{8}-\dfrac{t+2}{2}$

8. _________________

9. $\dfrac{3r+t}{5r^2t}-\dfrac{5r-4t}{rt^2}$

9. _________________

10. $\dfrac{8}{x+1}-\dfrac{5}{x-1}$

10. _________________

11. $\dfrac{4}{x^2+x-20}-\dfrac{3}{x^2-25}$

11. _________________

12. $\dfrac{2a-5}{15a}-\dfrac{7a-1}{10a}$

12. _________________

13. $\dfrac{3}{x}-\dfrac{9}{-x}$

13. _________________

14. $\dfrac{5-t}{t-4}-\dfrac{3t-2}{4-t}$

14. _________________

15. $\dfrac{x-3}{x^2-16}-\dfrac{5-x}{16-x^2}$

15. _________________

176

16. $\dfrac{3a}{a^2-1}-\dfrac{2}{1-a}$

16. _______________

17. $\dfrac{5}{4y^2-4y}-\dfrac{3}{4y-4}$

17. _______________

18. $\dfrac{x}{x^2+4x+3}-\dfrac{1}{x^2-1}$

18. _______________

Objective b   Simplify combined additions and subtractions of rational expressions.

Perform the indicated operations and simplify.

19. $\dfrac{2(3t+1)}{t-2}-\dfrac{5(2t-3)}{2-t}+\dfrac{7t+6}{t-2}$

19. _______________

20. $\dfrac{2a-b}{a-b}+\dfrac{a+2b}{b-a}-\dfrac{2a}{a-b}$

20. _______________

177

21. $\dfrac{8}{3x-1}-\dfrac{4}{1-3x}+\dfrac{2x}{3x-1}+\dfrac{x-5}{1-3x}$

21. _______________

22. $\dfrac{x+5}{x-3}-\dfrac{2-x}{x+3}-\dfrac{4x-18}{9-x^2}$

22. _______________

23. $\dfrac{3x}{1-4x}+\dfrac{2x}{4x+1}-\dfrac{1}{16x^2-1}$

23. _______________

24. $\dfrac{1}{a-b}+\dfrac{2}{a-b}+\dfrac{2a}{a^2-b^2}$

24. _______________

178

Chapter 6 RATIONAL EXPRESSIONS AND EQUATIONS

6.6 Complex Rational Expressions

Learning Objective
a   Simplify complex rational expressions.

Objective a   Simplify complex rational expressions.

Simplify.

1. $\dfrac{1+\dfrac{7}{12}}{1-\dfrac{2}{3}}$

1. ______________

2. $\dfrac{1-\dfrac{3}{8}}{1+\dfrac{7}{8}}$

2. ______________

3. $\dfrac{\dfrac{1}{4}+\dfrac{2}{5}}{\dfrac{3}{10}-\dfrac{4}{5}}$

3. ______________

4. $\dfrac{\dfrac{1}{a}+2}{\dfrac{1}{a}-4}$

4. _______________

5. $\dfrac{9-\dfrac{1}{t^2}}{3-\dfrac{1}{t}}$

5. _______________

6. $\dfrac{6+\dfrac{6}{n}}{2+\dfrac{2}{n}}$

6. _______________

7. $\dfrac{\dfrac{x}{10}-\dfrac{10}{x}}{\dfrac{1}{x}+\dfrac{1}{10}}$

7. _______________

8. $\dfrac{\dfrac{1}{p}+1}{\dfrac{1}{p^2}-1}$

8. _______________

180

9. $\dfrac{\dfrac{1}{2}-\dfrac{1}{x}}{\dfrac{2-x}{2}}$

10. $\dfrac{\dfrac{3}{y^2}-\dfrac{3}{z^2}}{\dfrac{1}{y}+\dfrac{1}{z}}$

11. $\dfrac{x-4-\dfrac{5}{x}}{x-2-\dfrac{3}{x}}$

12. $\dfrac{\dfrac{7}{a^3}-\dfrac{2}{a^2}}{\dfrac{5}{a^2}+\dfrac{4}{a}}$

9. _______________

10. _______________

11. _______________

12. _______________

13. $\dfrac{\dfrac{3}{5n^5} - \dfrac{1}{15n}}{\dfrac{6}{7n^3} + \dfrac{2}{21n}}$

13. _______________

14. $\dfrac{\dfrac{x}{y} + \dfrac{w}{z}}{\dfrac{y}{x} + \dfrac{z}{w}}$

14. _______________

15. $\dfrac{\dfrac{a}{4b^2} + \dfrac{5}{12b}}{\dfrac{5}{12b} + \dfrac{a}{4b^2}}$

15. _______________

16. $\dfrac{\dfrac{5}{n+1} + \dfrac{2}{n}}{\dfrac{3}{n+1} + \dfrac{5}{n}}$

16. _______________

182

Chapter 6 RATIONAL EXPRESSIONS AND EQUATIONS

6.7 Solving Rational Equations

Learning Objective
a   Solve rational equations.

Key Terms
Use the vocabulary terms listed below to complete each statement in Exercises 1–4.

check	clear	LCM	rational

1. The equation $\dfrac{2}{x} = \dfrac{3}{x+1}$ is an example of a(n) _________________ equation.

2. To solve a rational equation, first _________________ the equation of fractions.

3. When solving a rational equation, multiply on both sides of the equation by the

    _________________ of all the denominators.

4. After solving a rational equation, always _________________ possible solutions in

    the original equation.

Objective a   Solve rational equations.

Solve. Don't forget to check!

5. $\dfrac{2}{5} - \dfrac{3}{4} = \dfrac{x}{8}$ 5. _________________

6. $\dfrac{2}{3}+\dfrac{3}{4}=\dfrac{1}{x}$

6. _________________

7. $\dfrac{5}{6}+\dfrac{2}{5}=\dfrac{x}{15}$

7. _________________

8. $\dfrac{1}{x}=\dfrac{5}{8}-\dfrac{7}{9}$

8. _________________

9. $\dfrac{1}{5}+\dfrac{1}{8}=\dfrac{1}{t}$

9. _________________

10. $x+\dfrac{3}{x}=4$

10. _________________

184

11. $\dfrac{x}{7} - \dfrac{7}{x} = 0$

11. _______________

12. $\dfrac{8}{y} = \dfrac{9}{y} - \dfrac{1}{12}$

12. _______________

13. $\dfrac{6}{5a} + \dfrac{2}{a} = 1$

13. _______________

14. $\dfrac{n-4}{n+1} = \dfrac{6}{11}$

14. _______________

15. $\dfrac{3}{x-1} = \dfrac{2}{x+2}$

15. _______________

16. $\dfrac{x}{4}-\dfrac{x}{6}=\dfrac{1}{4}$

16. ____________________

17. $\dfrac{a+3}{8}-\dfrac{a-3}{6}=1$

17. ____________________

18. $\dfrac{7}{a+3}=\dfrac{4}{a-4}$

18. ____________________

19. $\dfrac{x-6}{2x+3}=\dfrac{1}{4}$

19. ____________________

20. $\dfrac{x-3}{x-8}=\dfrac{5}{x-8}$

20. ____________________

186

21. $\dfrac{3}{x+5} = \dfrac{7}{x}$

21. _______________

22. $\dfrac{x+3}{x+9} = \dfrac{x-2}{x+1}$

22. _______________

23. $\dfrac{1}{x+1} + \dfrac{1}{x-1} = \dfrac{1}{x^2-1}$

23. _______________

24. $\dfrac{3}{a+4} - \dfrac{10}{a^2-16} = 1$

24. _______________

25. $\dfrac{5-x}{7-x} = \dfrac{2}{x-7}$

25. _______________

26. $2 - \dfrac{x-3}{x+5} = \dfrac{x^2-17}{x+5}$

26. _______________

27. $3x - 1 = \dfrac{4x}{x+2}$

27. _______________

28. $\dfrac{4}{x-3} = \dfrac{2x}{x^2-9} - \dfrac{7}{x+3}$

28. _______________

188

Chapter 6 RATIONAL EXPRESSIONS AND EQUATIONS

6.8 Applications Using Rational Equations and Proportions

Learning Objectives
a   Solve applied problems using rational equations.
b   Solve proportion problems.

Key Terms
Use the vocabulary terms listed below to complete each statement in Exercises 1–4.

    proportion   **ratio**   **rate**   **similar**

1.   A(n) _________________ of two quantities is their quotient.

2.   The ratio of two different kinds of measure is called a(n) _________________.

3.   The equation $\dfrac{A}{B} = \dfrac{C}{D}$ is an example of a(n) _________________.

4.   Two triangles are _________________ if their corresponding angles have the same measure.

Objective a   Solve applied problems using rational equations.

Solve.

5.   It takes Alexis 3 hr to clean her family's garage. Alyssa   5. _________________
     takes 4 hr to do the same job. How long would it take
     them, working together, to clean the garage?

6.   Ethan can weed the flowerbeds by his office in 50 min.   6. _________________
     Anthony can do the same job in 45 min. How long would
     it take Ethan and Anthony to weed the beds if they worked
     together?

189

7. Abigail can file a week's worth of invoices in 75 min. Ava can do the same job in 90 min. How long would it take if they worked together?

7. _______________

8. Around Town Services owns two copy machines, an HP Color LaserJet 2800 and an HP Color LaserJet CM1015. The 2800 can copy a month's worth of advertising fliers in 5 hr. The CM1015 can do the same job in 12 hr. How long would the copiers take to copy the fliers if they work together?

8. _______________

9. Maya drives 20 km/h faster than Tara. While Tara travels 180 km, Maya travels 260 km. Find their speeds.

9. _______________

10. The speed of a freight train is 15 mph slower than the speed of a passenger train. The freight train travels 180 mi in the same time that it takes a passenger train to travel 240 mi. Find the speed of each train.

10. _______________

11. The speed of Tom's scooter is 16 mph less than the speed of Mary Lynn's motorcycle. The motorcycle can travel 290 mi in the same time that the scooter can travel 210 mi. Find of the speed of each vehicle.

11. _______________

190

12. Brody rides his bicycle at the same speed that Aeron rides his. It takes Brody $\frac{1}{3}$ hr more than it takes Aeron to ride to work. If Brody is 25 mi from work and Aeron is 20 mi from work, how long does it take Aeron to ride to work?

12. _________________

Objective b   Solve proportion problems.

Find the ratio of the following. Simplify if possible.

13. 360 mi, 15 gal

13. _________________

14. 340 km, 4 hr

14. _________________

Solve.

15. Approximately 100 cocoa beans are required to make $\frac{1}{4}$ lb of chocolate. How many beans are required to make $2\frac{1}{2}$ lb of chocolate?

15. _________________

16. Teri wrote 72 pages for her novel over a period of 12 days.   **16.** _________________
At this rate, how many pages would she write in 16 days?

17. Linda walked 610 steps in 5 min on an elliptical trainer. At   **17.** _________________
this rate, how many steps would she walk in 12 min?

18. The ratio of buttermilk to whole wheat flour in a flat bread   **18.** _________________
recipe is $\frac{2}{3}$. If 3 cups of buttermilk are used, how many
cups of whole wheat flour are used?

19. To determine the number of trout in his pond, Oak catches   **19.** _________________
25 trout, tags them, and lets them loose. Later, he catches
18 trout; 10 of them have tags. Estimate the number of
trout in the pond.

192

20. A sample of 48 memory cards contained 3 defective cards. **20.** _________________
How many defective cards would you expect in a sample
of 192 cards?

For each pair of similar triangles, find the length of the indicated side.

21. x **21.** _________________

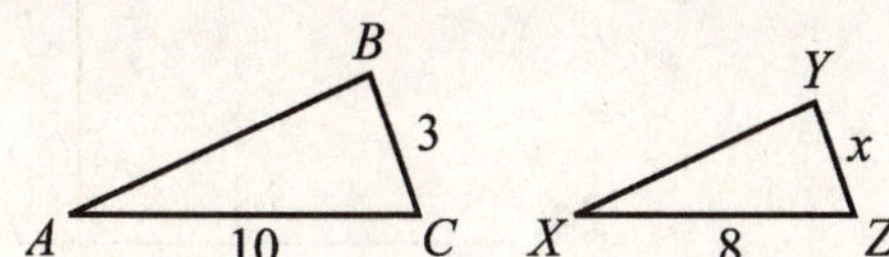

22. p **22.** _________________

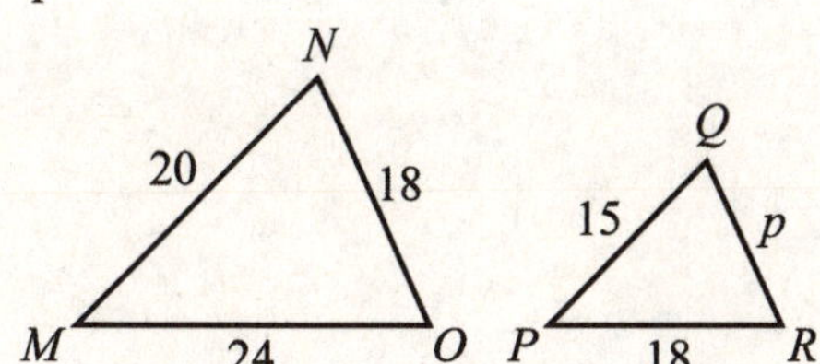

23. *g*

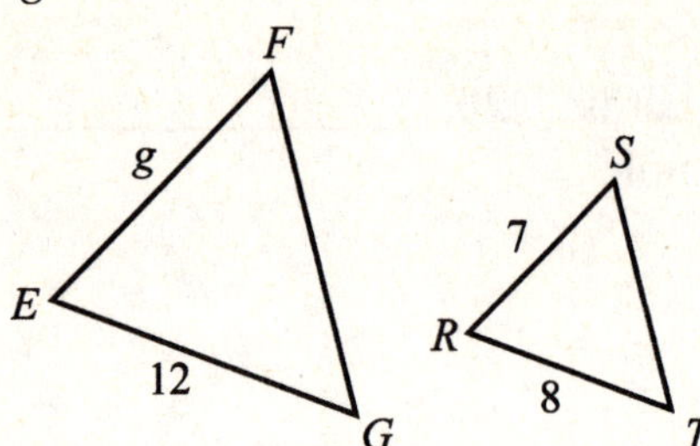

23. ______________

24. *y*

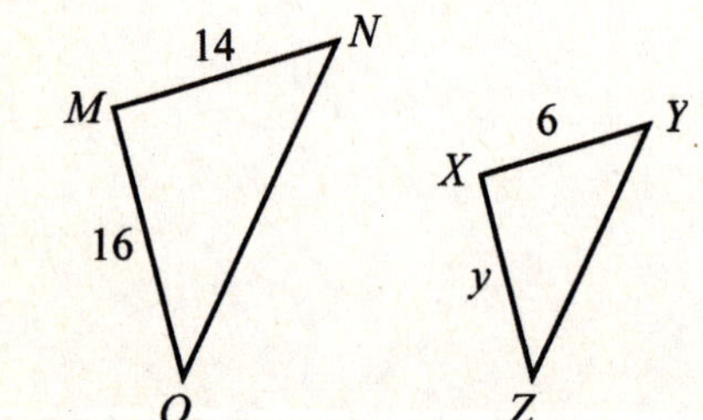

24. ______________

Chapter 6 RATIONAL EXPRESSIONS AND EQUATIONS

6.9 Variation and Applications

Learning Objectives
a   Find an equation of direct variation given a pair of values of the variables.
b   Solve applied problems involving direct variation.
c   Find an equation of inverse variation given a pair of values of the variables.
d   Solve applied problems involving inverse variation.
e   Find equations of other kinds of variation given values of the variables.
f   Solve applied problems involving other kinds of variation.

Key Terms
Use the vocabulary terms listed below to complete each statement in Exercises 1–4.

    direct   **inverse**   **proportionality**   **variation**

1.   The equation $y = kx$ is called an equation of _____________________ variation.

2.   In the equation $y = kx$, k is called the constant of _____________________.

3.   The equation $y = k / x$ is called an equation of _____________________ variation.

4.   In the equation $y = k / x$, k is called the _____________________ constant.

Objective a   Find an equation of direct variation given a pair of values of the variables.

Find the variation constant and an equation of variation in which y varies directly as x and the following are true.

5. $y = 55$ when $x = 5$ 5. _________________

6. $y = \dfrac{2}{3}$ when $x = 8$ 6. _________________

Objective b   Solve applied problems involving direct variation.

Solve.

7.   Hooke's law states that the distance d that a spring
is stretched by a hanging object varies directly as
the mass m of the object. If the distance is 50 cm
when the mass is 6 kg, what is the distance when
the mass is 2 kg?

7. _______________

8.   The electric current I, in amperes, in a circuit varies
directly as the voltage V. When 12 volts are applied,
the current is 3 amperes. What is the current when
16 volts are applied?

8. _______________

9.   The number of calories burned by a person in a
Zumba aerobic class is directly proportional to
the time spent exercising. It takes 10 min to burn
80 calories (*Source*: Family Fun and Fitness).
How long would it take to burn 200 calories in
the class?

9. _______________

Objective c   Find an equation of inverse variation given a pair of values of the
variables.

*Find the variation constant and an equation of variation in which y varies inversely as
x and the following are true.*

10. $y = 7$ when $x = 9$

10. _______________

196

11. $y = 72$ when $x = \dfrac{1}{12}$

11. ______________

Objective d   Solve applied problems involving inverse variation.

Solve.

12. The time T required to do a job varies inversely as the number of people P working. It takes 4 hr for 9 people to weed the community garden. How long would it take 10 people to complete the job?

12. ______________

13. The wavelength W of a radio wave varies inversely as its frequency F. A wave with a frequency of 1600 kilohertz has a length of 225 meters. What is the length of a wave with a frequency of 3000 kilohertz?

13. ______________

14. The current I in an electrical conductor varies inversely as the resistance R in the conductor. If the current is $\frac{2}{5}$ ampere when the resistance is 120 ohms, what is the current when the resistance is 150 ohms?

14. ______________

Objective e   Find equations of other kinds of variation given values of the variables.

Find an equation of variation in which the following are true.

15. y varies inversely as the square of x, and $y = 72$ when $x = 0.5$.

15. ______________

197

16. y varies directly as the square of x, and $y = 0.44$ when $x = 0.1$.

16. _________ _________

17. y varies jointly as x and the square of z, and $y = 385$ when $x = 14$ and $z = 5$.

17. _________________

18. y varies jointly as w and the square of x and inversely as z, and $y = 9$ when $w = 3$, $x = 3$ and $z = 15$.

18. _________________

Objective f   Solve applied problems involving other kinds of variation.

Solve.

19. The intensity I of a wireless signal varies inversely as the square of the distance d from the transmitter. If the intensity is 50 W/m^2 at a distance of 5 km, what is the intensity 20 km from the transmitter?

19. _________________

20. Atmospheric drag varies jointly as an object's surface area A and its velocity v. If a car traveling at a velocity of 80 mph with a surface area of 41.2 sq ft experiences a drag of 170 N, how fast must a car with 42.8 sq ft of surface area go to experience 409 N of drag? Round to the nearest integer.

20. _________________

198

Chapter 7 GRAPHS, FUNCTIONS AND APPLICATIONS

7.1 Functions and Graphs

> **Learning Objective**
> a Determine whether a correspondence is a function.
> b Given a function described by an equation, find function values (outputs) for specified values (inputs).
> c Draw the graph of a function.
> d Determine whether a graph is that of a function using the vertical-line test.
> e Solve applied problems involving functions and their graphs.

Key Terms
Use the vocabulary terms listed below to complete each statement in Exercises 1–5.

   function   input   independent   relation   outputs

1. A(n) ___________________ is a member of the domain of a function.

2. A ___________________ is a rule that assigns to each member of some set exactly one member of another set.

3. Function values are also called ___________________.

4. The variable x in $f(x) = 2x - 7$ is called the ___________________ variable.

5. A ___________________ is a rule that assigns to each member of some set at least one member of another set.

Objective a   Determine whether a correspondence is a function.

Determine whether each correspondence is a function.

6. 6. ___________________

$$-5 \longrightarrow -4$$
$$-3 \longrightarrow 0$$
$$-1 \longrightarrow 1$$
$$2 \longrightarrow 5$$
$$3 \nearrow$$

7. 7. ___________________

$$-5 \longrightarrow -4$$
$$-3 \longrightarrow 0$$
$$-1 \longrightarrow 3$$
$$2 \longrightarrow 5$$

8. Determine if the relation is a function.
   Domain: Each person in a town
   Correspondence: Each person's uncle
   Range: A set of males

8. _______________

Objective b   Given a function described by an equation, find function values (outputs) for specified values (inputs).

Find the function values.

9. $g(v) = 7v + 1$; find $g(4)$, $g(-2)$, and $g(6.5)$.

9. _______________

10. $g(x) = -2x - 7$; find $g(-2)$.

10. _______________

11. $f(x) = 3x^2 - 3x$; find $f(0)$, $f(-1)$, and $f(2)$.

11. _______________

12. As the price of a product increases, the consumer's purchases, or demand, for the product decreases. Suppose that under certain conditions in our economy, the demand for sugar is related to price by the demand function $d(p) = -1.6p + 24.1$, where p is the price of a 5-lb bag of sugar and $d(p)$ is the quantity of 5-lb bags, in millions, purchased at price p. What is the quantity purchased when the price is \$1 per 5-lb bag? \$2 per 5-lb bag?

12. _______________

200

Objective c   Draw the graph of a function.

Graph each function.

13. $f(x) = 3x + 1$

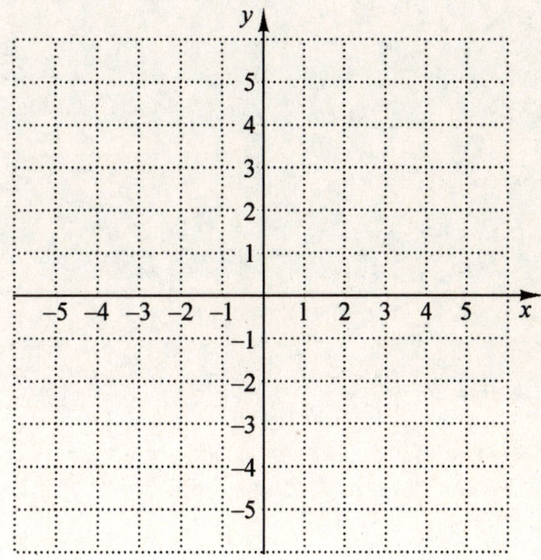

14. $f(x) = -\dfrac{2}{3}x + 3$

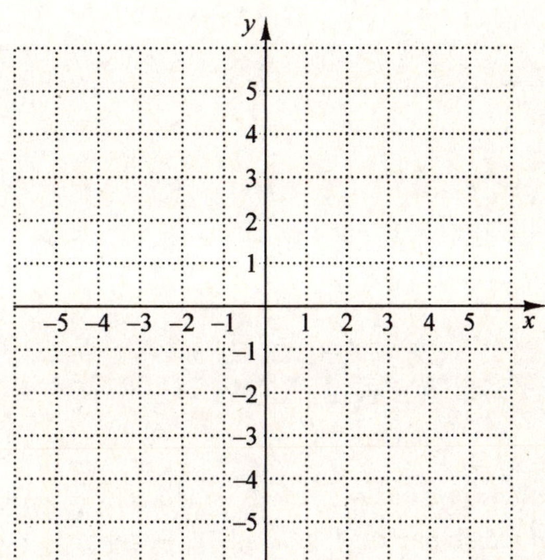

15. $g(x) = 4|x|$

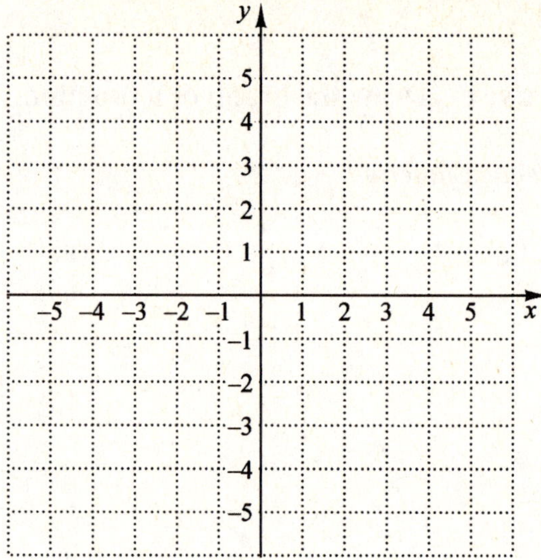

16. $f(x) = 3 - x^2$

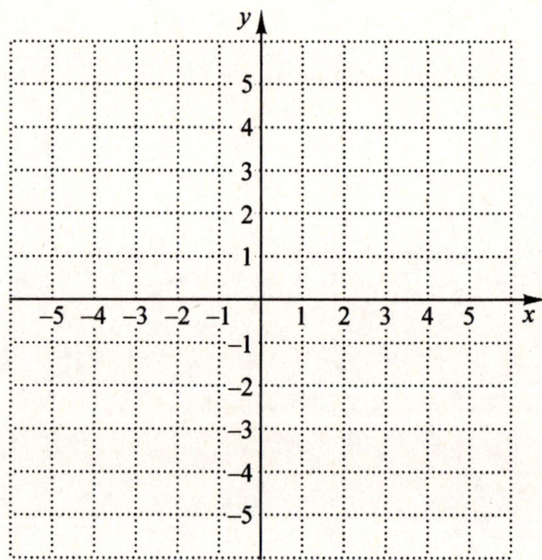

17. $f(x) = x^2 - 4x + 4$

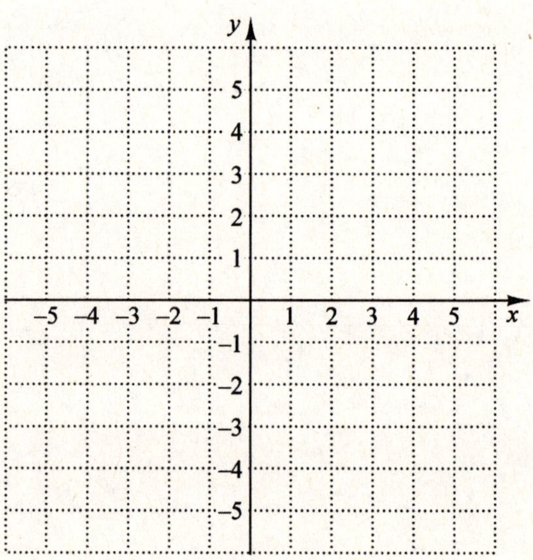

202

Objective d   Determine whether a graph is that of a function using the vertical-line test.

Determine whether each of the following is the graph of a function.

18.

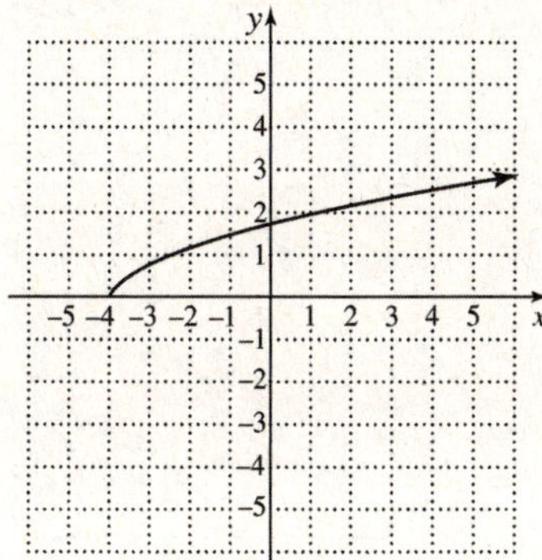

18. _______________

19.

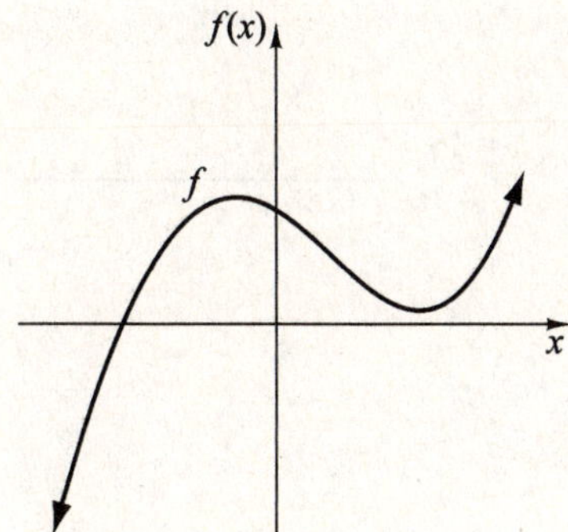

19. _______________

20.

20. _______________

21.

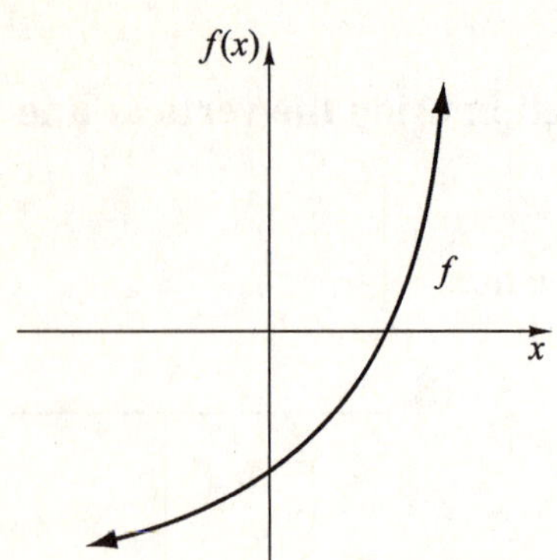

21. _______________

Objective e   Solve applied problems involving functions and their graphs.

The following graph approximates the number of words participants were able to memorize. The number of words is a function f of the time, in minutes.

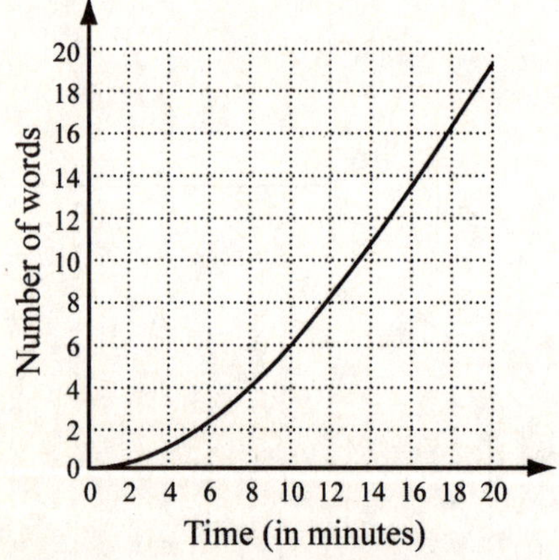

22. Approximate the number of words memorized after 10 minutes. That is, find $f(10)$.

22. _______________

23. Approximate the number of words memorized after 18 minutes. That is, find $f(18)$.

23. _______________

204

Chapter 7 GRAPHS, FUNCTIONS AND APPLICATIONS

7.2 Finding Domain and Range

Learning Objectives
a   Find the domain and the range of a function.

Key Terms
Use the vocabulary terms listed below to complete each statement in Exercises 1–4.

domain	**function**	**relation**	**range**

1. The _____________________ is the set of all first coordinates.

2. The _____________________ is the set of all second coordinates.

3. A set of ordered pairs is called a _____________________ .

4. When a set of ordered pairs is such that no two different pairs share a common first

    coordinate, we have a _____________________ .

Objective a   Find the domain and the range of a function.

In Exercises 5-7, the graph is that of a function. Determine for each one **(a)** $f(1)$; **(b)** *the domain;* **(c)** *any x-values for which* $f(x) = 2$; *and* **(d)** *the range.*

5.

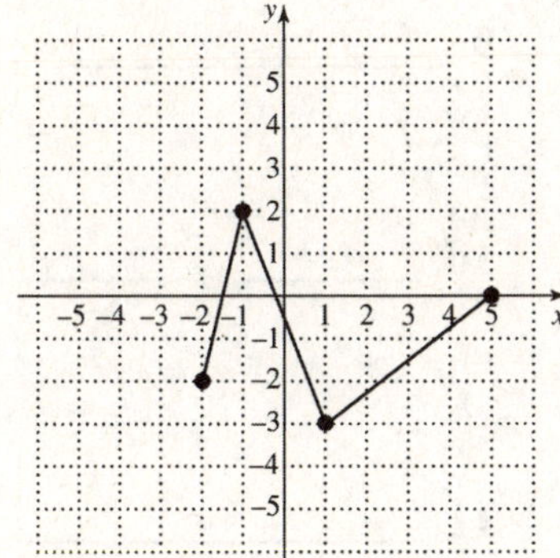

5.

a._____________________

b._____________________

c._____________________

d._____________________

6.

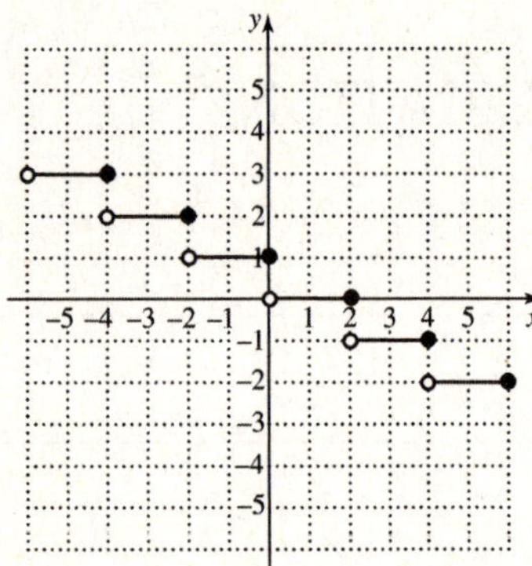

7.

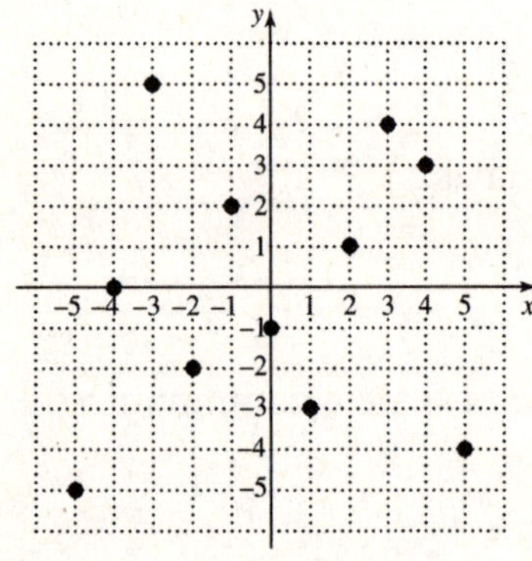

Find the domain.

8. $f(x) = \dfrac{7}{x-1}$

8. ________________

9. $f(x) = |5x + 4|$

9. ________________

10. $g(x) = \dfrac{4}{5}x$

10. ________________

11. $g(x) = \dfrac{3}{|7x-5|}$

11. ________________

206

Chapter 7 GRAPHS, FUNCTIONS AND APPLICATIONS

7.3 Linear Functions: Graphs and Slope

Learning Objectives
a   Find the y-intercept of a line from the equation $y = mx + b$ or $f(x) = mx + b$.
b   Given two points on a line, find the slope. Given a linear equation, derive the equivalent slope-intercept equation and determine the slope and the y-intercept.
c   Solve applied problems involving slope.

Key Terms
Use the vocabulary terms listed below to complete each statement in Exercises 1–7.

run	**rise**	**grade**	**up**	**down**	*m*	*b*

1.   The slope of any line written in the form $y = mx + b$ is _______________________________.

2.   The y-intercept of any line written in the form $y = mx + b$ is _____________________________.

3.   Lines with positive slope slant ______________________ from left to right.

4.   Lines with negative slope slant ______________________ from left to right.

5.   The change in y as we move from one point to another is called the ________________ .

6.   The change in x as we move from one point to another is called the ________________ .

7.   The ____________________ of the road is a measure of how steep it is.

Objective a   Find the y-intercept of a line from the equation $y = mx + b$ or $f(x) = mx + b$.
Objective b   Given two points on a line, find the slope. Given a linear equation, derive the equivalent slope-intercept equation and determine the slope and the y-intercept.

Find the slope and the y-intercept.

8. $y = -\dfrac{3}{5}x + 2$ 8. ________________

9. $4x - 5y = 9$ 9. ________________

10. $y = \dfrac{4}{9}x - 5$

10. ___________________

11. $2x + 3y = 9$

11. ___________________

12. $g(x) = -3.5x$

12. ___________________

13. $5x - 3y = -9$

13. ___________________

Objective b   Given two points on a line, find the slope. Given a linear equation, derive the equivalent slope-intercept equation and determine the slope and the *y*-intercept.

Find the slope of each line.

14.

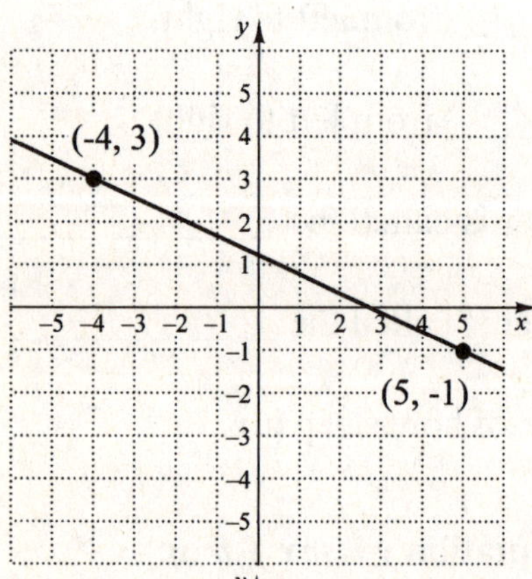

14. ___________________

15.

15. ___________________

Find the slope of the line containing the given pair of points.

16. $\left(3,-\dfrac{1}{3}\right),\ \left(-2,\ \dfrac{5}{3}\right)$

16. _________________

17. $\left(\dfrac{1}{2},-2\right),\ \left(-\dfrac{1}{2},-5\right)$

17. _________________

18. $(0.6,-0.8),\ (0.1,-0.8)$

18. _________________

19. $(10,\ 15),\ (-25,-8)$

19. _________________

Objective c   Solve applied problems involving slope.

Find the slope (or rate of change).

20. A road rises 7 ft over a horizontal distance of 280 ft. Find the slope (or grade) of the road.

20. _________________

21. A ramp rises 3 ft over a horizontal distance of 50 ft. Find the grade of the ramp.

21. _________________

Find the rate of change.

22.

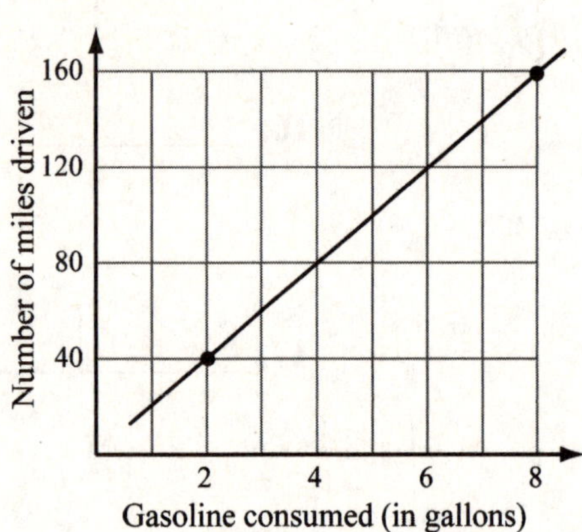

22. ________________

23.

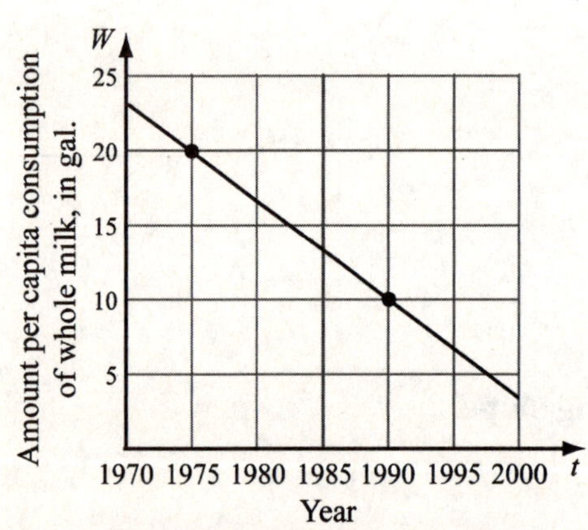

23. ________________

24.

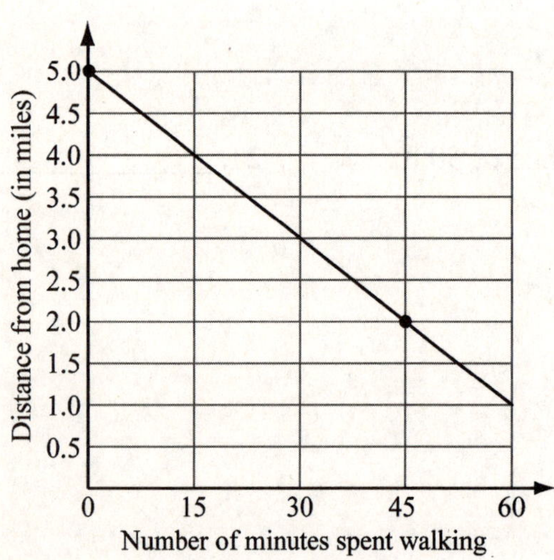

24. ________________

210

Chapter 7 GRAPHS, FUNCTIONS AND APPLICATIONS

7.4 More on Graphing Linear Equations

Learning Objective
a   Graph linear equations using intercepts.
b   Given a linear equation in slope-intercept form, use the slope and the y-intercept to graph the line.
c   Graph linear equations of the form $x = a$ or $y = b$.
d   Given the equations of two lines, determine whether their graphs are parallel or whether they are perpendicular.

Key Terms
Use the vocabulary terms listed below to complete each statement in Exercises 1–8.

not defined	x-intercept	y-intercept	0
horizontal	parallel	perpendicular	vertical

1. Two nonvertical lines are ___________________ if they have the same slope and different y-intercepts.

2. Two nonvertical lines are ___________________ if the product of their slopes is -1.

3. Parallel ___________________ lines have equations $x = p$ and $x = q$, where $p \neq q$.

4. If one equation in a pair of perpendicular lines is vertical, then the other is ___________________.

5. The ___________________ occurs where a graph crosses the x-axis.

6. The ___________________ occurs where a graph crosses the y-axis.

7. The slope of a horizontal line is ___________________.

8. The slope of a vertical line is ___________________.

Objective a   Graph linear equations using intercepts.

Find the intercepts and then graph the line.

9. $x + 2y = 4$

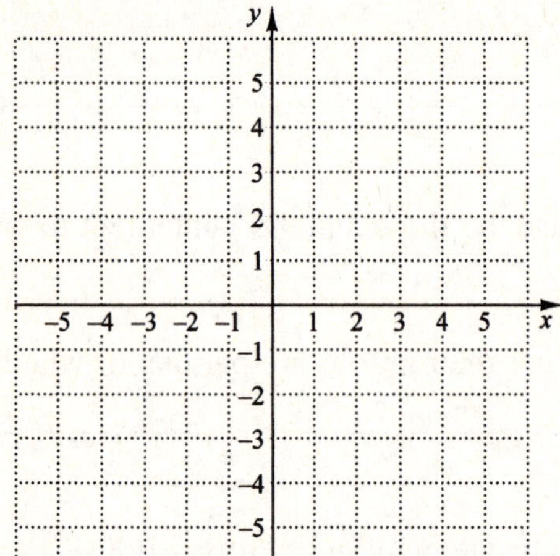

9. _______________

10. $3y - 3 = x$

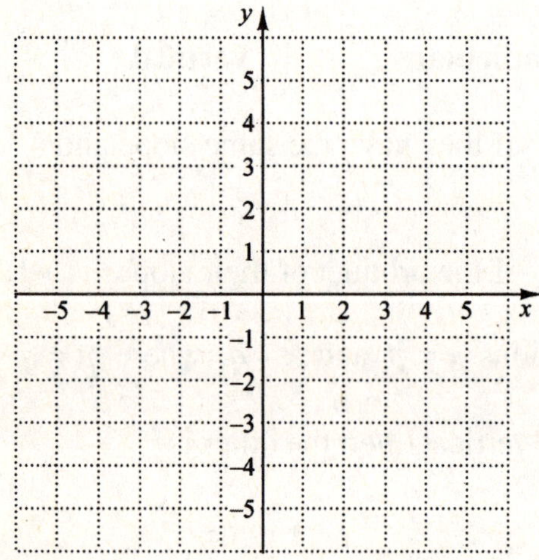

10. _______________

11. $x + 5 = y$

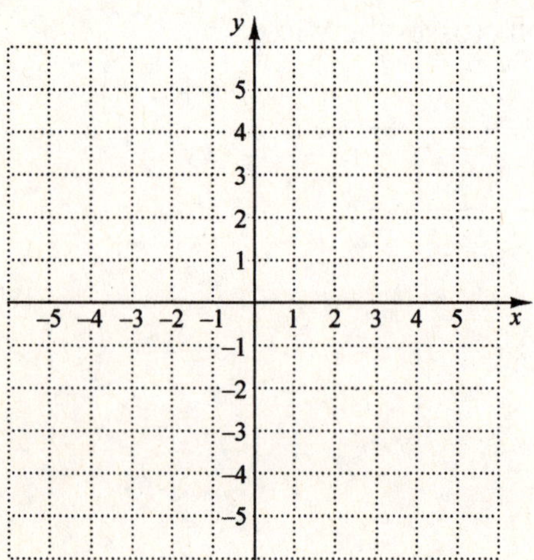

11. _______________

212

12. $4x - 5y = 20$ 12. _______________

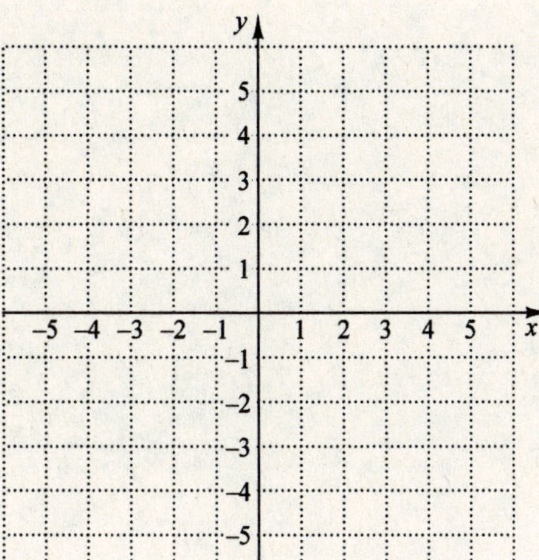

Objective b   **Given a linear equation in slope-intercept form, use the slope and the y-intercept to graph the line.**

Graph using the slope and the y-intercept.

13. $4x + 2y = 8$ 13. _______________

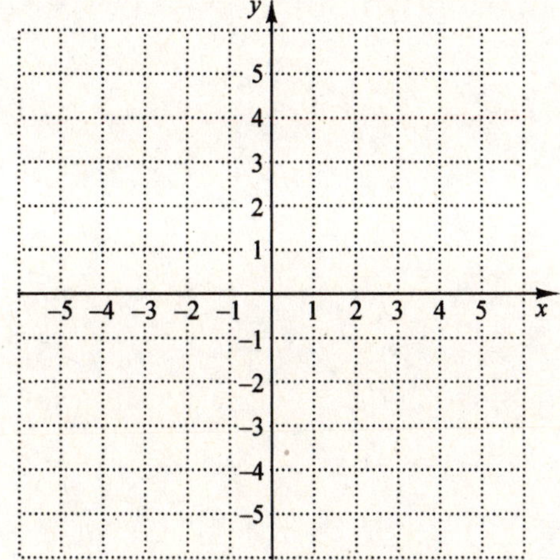

14. $y = -4 - 4x$

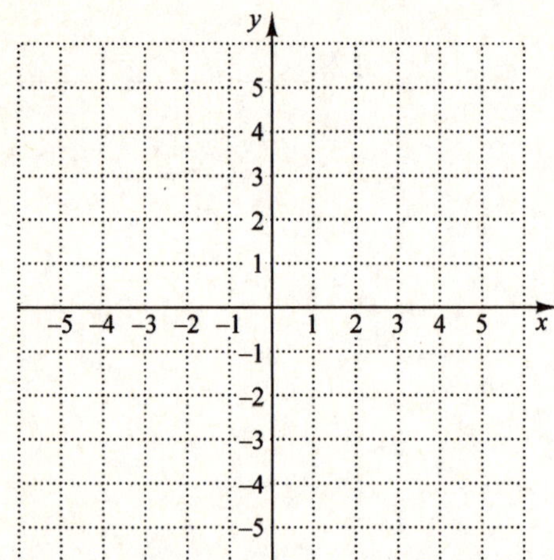

14. ______________

Objective c   Graph linear equations of the form $x = a$ or $y = b$.

Graph and, if possible, determine the slope.

15. $x = -3$

15. ______________

214

16. $y = 4$

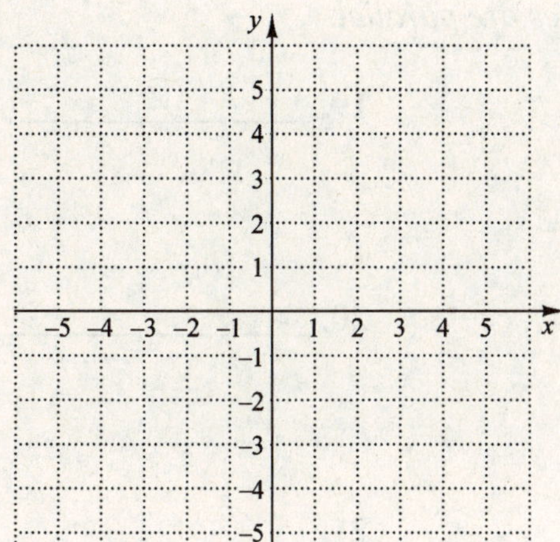

16. 

17. $2y = -5$

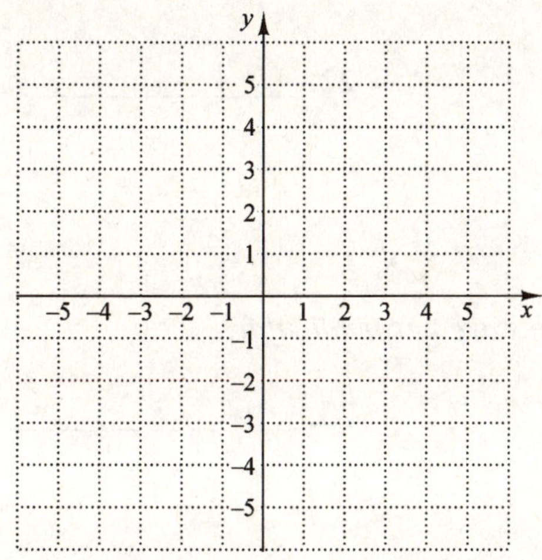

17. 

18. $3x - 15 = 0$

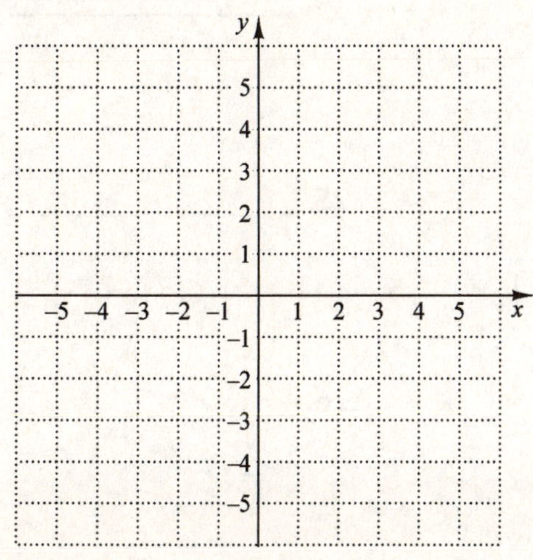

18. _______________

Objective d   Given the equations of two lines, determine whether their graphs are parallel or whether they are perpendicular.

Determine whether the graphs of the given pair of lines are parallel.

19. $x - y = 7,$
$\quad y = x + 3$

19. ________________

20. $y = 2x - 3,$
$\quad 4x - 2y = 1$

20. ________________

21. $y + 2 = 5x,$
$\quad 5x + y = 3$

21. ________________

22. $y = 3,$
$\quad y = -5$

22. ________________

Determine whether the graphs of the given pair of lines are perpendicular.

23. $y = 5 - 2x,$
$\quad 2y + x = 1$

23. ________________

24. $x - y = 3,$
$\quad x = y + 5$

24. ________________

25. $y = x + 1,$
$\quad x + y = 4$

25. ________________

26. $x = 7,$
$\quad y = 0$

26. ________________

216

Chapter 7 GRAPHS, FUNCTIONS AND APPLICATIONS

7.5 Finding Equations of Lines; Applications

Learning Objective
a   Find an equation of a line when the slope and the y-intercept are given.
b   Find an equation of a line when the slope and a point are given.
c   Find an equation of a line when two points are given.
d   Given a line and a point not on the given line, find an equation of the line parallel to the line and containing the point, and find an equation of the line perpendicular to the line and containing the point.
e   Solve applied problems involving linear functions.

Key Terms

Use the vocabulary terms listed below to complete each statement in Exercises 1–6.

slope	**point-slope**	**slope-intercept**
y-intercept	**parallel**	**perpendicular**

1. The _________________ form of a line is written $y - y_1 = m(x - x_1)$.

2. Two lines with different y-intercepts are _________________ if they have the same slope.

3. Two lines are _________________ if the product of their slopes is -1.

4. The equation $y = mx + b$ is in _________________ form.

5. The graph of an equation $y = mx + b$ has _________________ m.

6. The graph of an equation $y = mx + b$ has _________________ $(0, b)$.

Objective a   **Find an equation of a line when the slope and the y-intercept are given.**

Find an equation of the line having the given slope and y-intercept.

7. Slope $= -15$, y-intercept $= (0, 12)$ 7. _________________

8. Slope $= 2.7$, y-intercept $= (0, -4.3)$ 8. _________________

Find a linear function f(x) = mx+b whose graph has the given slope and y-intercept.

9. Slope $= -\dfrac{5}{6}$, y-intercept $= (0, -8)$

9. _______________

10. Slope $= \dfrac{3}{4}$, y-intercept $= \left(0, -\dfrac{5}{11}\right)$

10. _______________

Objective b   Find an equation of a line when the slope and a point are given.

Find an equation of the line having the given slope and containing the given point.

11. $(-4, 0)$, $m = -3$

11. _______________

12. $(1, 6)$, $m = \dfrac{2}{3}$

12. _______________

13. $(5, -1)$, $m = 2$

13. _______________

14. $(0, 7)$, $m = -5$

14. _______________

Objective c   Find an equation of a line when two points are given.

Find an equation of the line containing the given pair of points.

15. $(3, 8)$ and $(1, 10)$

15. _______________

16. $(0, 3)$ and $(5, 2)$

16. _______________

218

17. $(4, 7)$ and $(-1, 3)$ 17. _________________

18. $(-7, 5)$ and $(-1, 3)$ 18. _________________

Objective d   Given a line and a point not on the given line, find an equation of the line parallel to the line and containing the point, and find an equation of the line perpendicular to the line and containing the point.

Write an equation of the line containing the given point and parallel to the given line.

19. $(2,-1)$; $4x-y=8$ 19. _________________

20. $(-12,0)$; $2x+3y=5$ 20. _________________

21. $(0,4)$; $y=8x-9$ 21. _________________

Write an equation of the line containing the given point and perpendicular to the given line.

22. $(-3,6)$; $5x+10y=2$ 22. _________________

23. $(0,8)$; $6x+7y=5$ 23. _________________

24. $(0,3)$; $y=x-7$ 24. _________________

219

Objective e   Solve applied problems involving linear functions.

Solve.

25. In 1992, 1250 students attended Eastside College.
By 2007, the college had 1425 students. Let $c(t)$
represent the number of students at the college t
years after 1992.
a. Find a linear function that fits the data.
b. Use the function of part (a) to predict the number
   of students in 2010.
c. When will 1600 students attend Eastside
   College?

25.
a._______________

b._______________

c._______________

26. In 1985, the record for the 100-m run at
Eastside College was 12.3 sec. In 2005, it
was 12.1 sec. Let $r(t)$ represent the record in
the 100-m run and t the number of years
since 1985.
a. Find a linear function that fits the data.
b. Use the function of part (a) to predict the record in
   2000 and in 2010.
c. Find the year when the record will be 12.0 sec.

26.
a._______________

b._______________

c._______________

220

Chapter 8 SYSTEMS OF EQUATIONS

8.1 Systems of Equations in Two Variables

Learning Objectives
a Solve a system of two linear equations or two functions by graphing and determine whether a system is consistent or inconsistent and whether the equations in a system are dependent or independent.

Key Terms
Use the vocabulary terms listed below to complete each statement in Exercises 1–6.

consistent   **dependent**   **inconsistent**

independent   **intersection**   **system**

1. A(n) ___________________ of equations is a set of two or more equations, in two or more variables, for which a common solution is sought.

2. When solving a system of equations graphically, we look for the ___________________ of the graphs of the equations.

3. A system of equations that has at least one solution is said to be ___________________.

4. A system of equations with no solution is called ___________________.

5. If a system of two equations in two variables has infinitely many solutions, the equations are ___________________.

6. If a system of two equations in two variables has one solution or no solutions, the equations are ___________________.

Objective a   Solve a system of two linear equations or two functions by graphing and determine whether a system is consistent or inconsistent and whether the equations in a system are dependent or independent.

Solve each system of equations graphically. Then classify the system as consistent or inconsistent and the equations as dependent or independent.

7. $x - y = 1,$
     $x + y = 5$

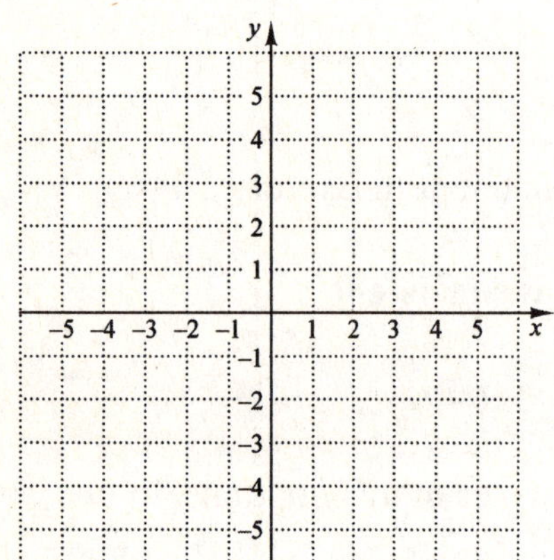

8. $6x - y = 1,$
     $3x + y = 8$

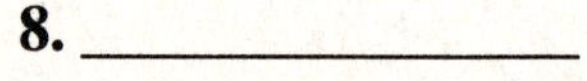

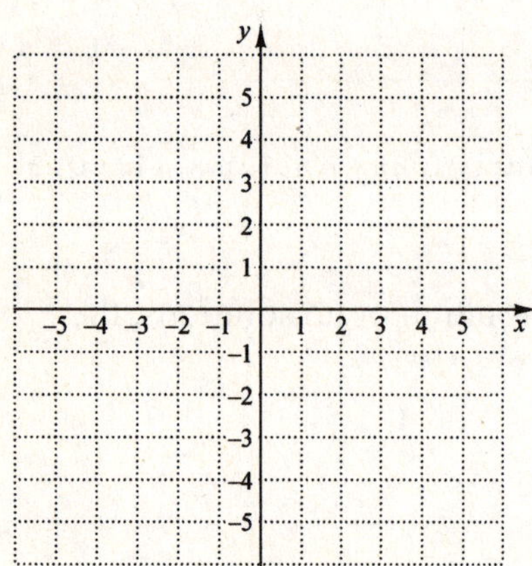

222

9. $y = \dfrac{1}{2}x,$

$x = 3y + 1$

9. _______________

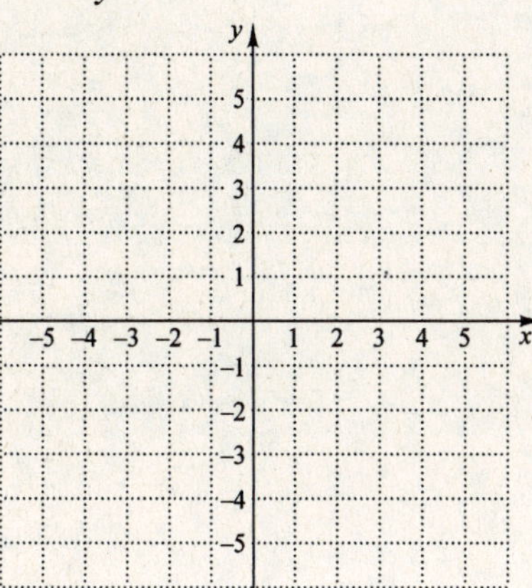

10. $x = 2y,$

$x - 2y = 2$

10. _______________

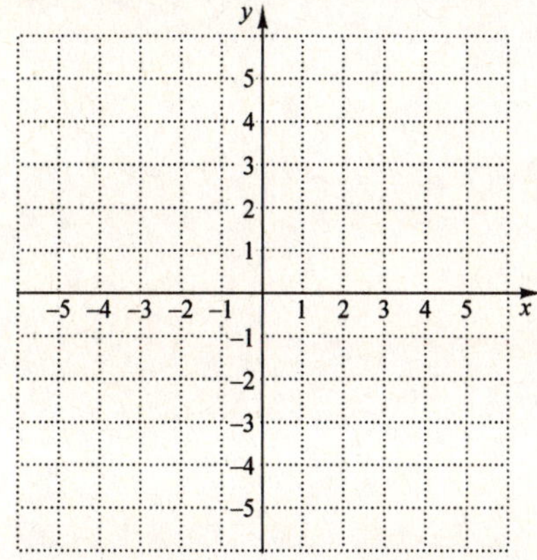

11. $x + y = 3,$

$x = 3 - y$

11. _______________

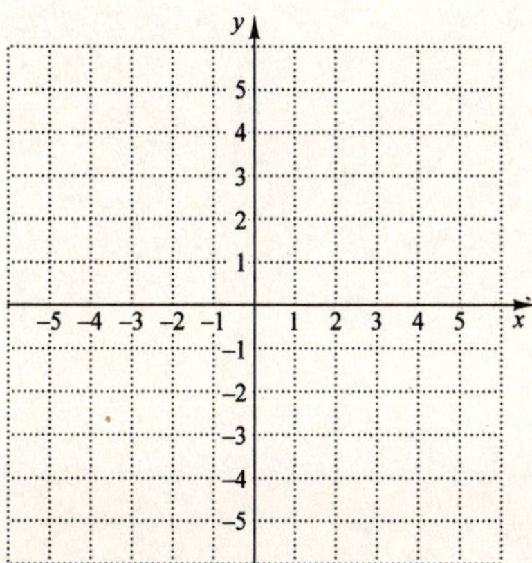

12. $x - \dfrac{1}{2}y = 3,$

$x = 3y - 2$

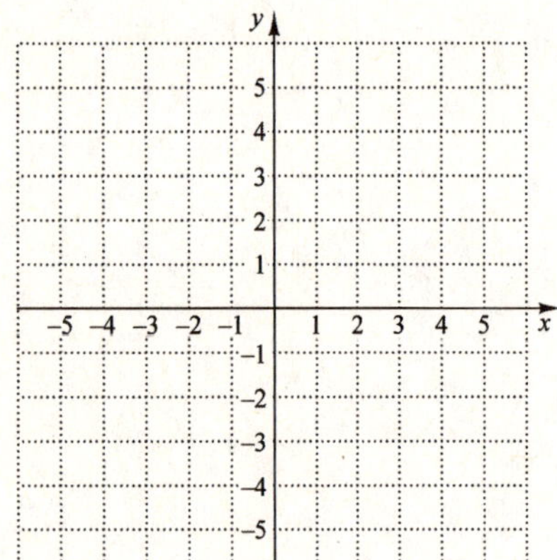

12. _______________

13. $2x - y = 2,$

$4x = 1 - y$

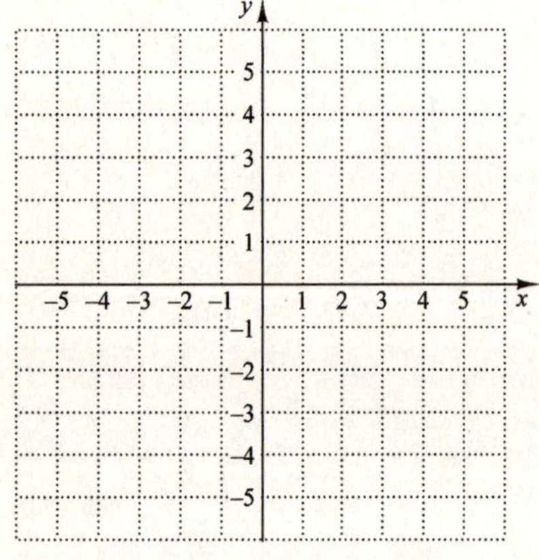

13. _______________

14. $x = -3,$

$y = 2$

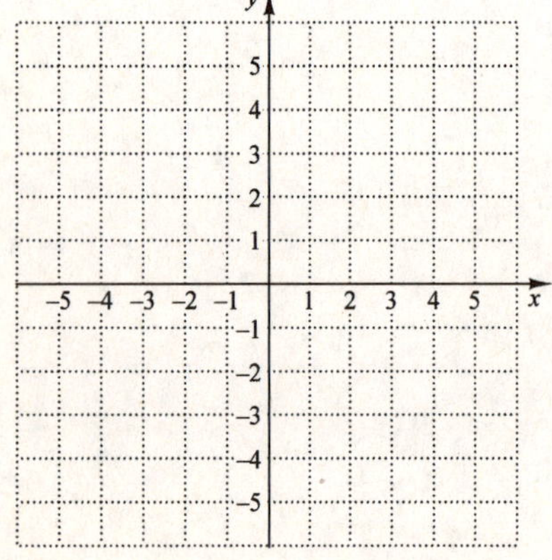

14. _______________

224

Chapter 8 SYSTEMS OF EQUATIONS

8.2 Solving by Substitution

Learning Objective
a   Solve systems of equations in two variables by the substitution method.
b   Solve applied problems by solving systems of two equations using substitution.

Key Terms

Use the vocabulary terms listed below to complete each statement in Exercises 1–4.

algebraic	**solve**	**substitute**	**translate**

1.   Nongraphical methods for solving systems of equations are called __________________ methods.

2.   One way to solve a system of two equations is to __________________ an equivalent expression for a variable into an equation.

3.   When using the substitution method, if neither equation of a pair has a variable alone on one side, we __________________ one equation for one of the variables.

4.   Sometimes it is easier to __________________ a problem into two equations in two variables.

Objective a   Solve systems of equations in two variables by the substitution method.

Solve each system of equations by the substitution method.

5. $x + y = 5$,   5.______________
     $y = x + 1$

6. $x = y + 6$,   6.______________
     $x + 2y = 12$

7. $y = 3x - 2$,
$2y - x = 1$

7.______________

8. $2x + 5y = 9$,
$x = y - 1$

8.______________

9. $a + b = -6$,
$b - a = 8$

9.______________

10. $3x - y = 3$,
$-5x + y = 3$

10.______________

11. $s - 6t = 5$,
$3s + 2t = 0$

11.______________

226

12. $3y + 4 = x$,
$\quad 3y - x = 5$

12.______________

13. $p - q = -5$,
$\quad 2p = q + 1$

13.______________

14. $x - 3y = 7$,
$\quad 3x - 5y = 1$

14.______________

15. $7x + 2y = -4$,
$\quad 8x - y = 2$

15.______________

Solve.

16. The perimeter of the largest possible regulation-size rugby field is 428 m. The length is 4 m longer than twice the width. Find the dimensions.

16. 

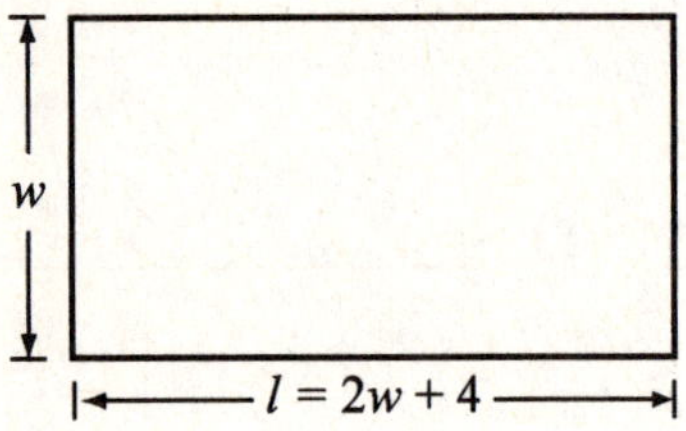

17. A rectangle has a perimeter of 124 ft. The width is 10 ft less than the length. Find the length and width.

17. 

18. The perimeter of a doubles-play badminton court is 128 ft. The width is 2 ft less than half the length. Find the length and the width.

18. ______________________

228

Chapter 8 SYSTEMS OF EQUATIONS

8.3 Solving by Elimination

Learning Objectives
a   Solve systems of equations in two variables by the elimination method.
b   Solve applied problems by solving systems of two equations using elimination.

Key Terms
Use the vocabulary terms listed below to complete each statement in Exercises 1–4.

elimination   **false**   **opposites**   **true**

1. The _____________________ method for solving systems of equations makes use of the

    addition principle.

2. To eliminate a variable when adding, the terms containing that variable must be

    _____________________.

3. When solving using the elimination method, obtaining a(n) _____________________

    equation means there is no solution.

4. When solving using the elimination method, obtaining a(n) _____________________

    equation means the system has an infinite number of solutions.

Objective a   Solve systems of equations in two variables by the elimination method.

Solve each system of equations by the elimination method.

5. $x - y = 3$,   5._____________________
    $x + y = 11$

6. $a + b = 8$,
$-a + 5b = 7$

6.________________

7. $3x - y = 3$,
$-5x + y = 3$

7.________________

8. $r + s = 9$,
$3r - 2s = -5$

8.________________

9. $8a + 12b = 10$,
$6a + 9b = 5$

9.________________

230

10. $3x - 2y = 5$,
$6x - 10 = 4y$

10._______________

11. $6x - 0.5y = 3$,
$1.5x + 2.25y = -4$

11._______________

12. $0.12s - 0.06t = 12$,
$0.08s + 0.16t = -16$

12._______________

Solve. Use the elimination method when solving the translated system.

13. In a recent basketball game, the Westside Wildcats scored 91 of their points on a combination of 41 two- and three-point baskets. How many of each type of shot were made?

13. 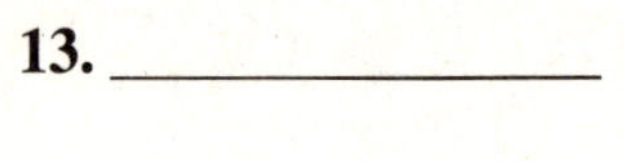

14. Supplementary angles are angles whose sum is $180°$. Two angles are supplementary. One angle measures $20°$ more than four times the measure of the other. Find the measure of each angle.

14. 

15. Complementary angles are angles whose sum is $90°$. Two angles are complementary. One angle is $10°$ less than four times the other. Find the measure of each angle.

15. ___________________

232

Chapter 8 SYSTEMS OF EQUATIONS

8.4 Solving Applied Problems: Two Equations

Learning Objectives
a   Solve applied problems involving total value and mixture using systems of two equations.
b   Solve applied problems involving motion using systems of two equations.

Key Terms

Use the vocabulary terms listed below to complete each numbered blank in the statement.

    rate   **motion**   **distance**   **time**

The __________________ formula states that __________________ =
            (1)   (2)

__________________ (or speed) · __________________ .
        (3)   (4)

1. __________________

2. __________________

3. __________________

4. __________________

Objective a   Solve applied problems involving total value and mixture using systems of two equations.

Solve.

5.   Admission to Mammoth Cave is $12 for adults and $8 for   **5.** ________________
youth (*Source:* National Park Service). One day, 575
people entered the cave, paying a total of $5600. How
many adults and how many youth entered the cave?

6. Alpine Trail Mix is 40% nuts and Meadows Trail Mix is
    25% nuts. How much of Alpine and how much of
    Meadows should be mixed to form a 10-lb batch of trail
    mix that is 32% nuts?

6. ______________________

7. Firehouse Cafe charges $6.40 for one bowl of soup and
    one vegetable tortilla wrap, and $16.60 for two bowls of
    soup and three wraps. Determine the cost of one bowl of
    soup and one vegetable tortilla wrap.

7. ______________________

8. The Clarkstown Volunteer Fire Department served 325
    chicken and noodle dinners. A child's plate cost $4.50 and
    an adult's plate cost $6.00. A total of $1672.50 was
    collected. How many of each type of plate was served?

8. ______________________

234

9. Tropical Punch is 18% fruit juice and Caribbean Spring is 24% fruit juice. How many liters of each should be mixed to get 18 L of a mixture that is 20% juice?

9. ______________

10. Tia received $7.55 in change in quarters and dimes. There were 38 coins in all. How many of each kind did she receive?

10. ______________

Objective b   Solve applied problems involving motion using systems of two equations.

Solve.

11. Two private airplanes leave an airport at the same time, both flying due east. One travels at 150 mph and the other travels at 165 mph. In how many hours will they be 45 mi apart?

11. ______________

12. A moving van leaves a rest stop on an interstate highway and travels north at 50 mph. An hour later, a car leaves the rest stop and travels north at 70 mph. When will the car catch up to the van?

12. ______________

13. Gavin's boat took 2 hr to make a trip downstream with a 4-mph current. The return trip against the same current took 3 hr. Find the speed of the boat in still water.

13. _______________

14. It takes a passenger train 1 hr less time than it takes a freight train to make the trip from Coastal City to Valley Village. The passenger train averages 72 km/h, while the freight train averages 54 km/h. How far is it from Coastal City to Valley Village?

14. _______________

15. Two cars leave at the same time and travel directly toward each other from points 100 mi apart at rates of 50 mph and 70 mph. When will they meet?

15. _______________

236

Chapter 8 SYSTEMS OF EQUATIONS

8.5 Systems of Equations in Three Variables

Learning Objective
a   Solve systems of three equations in three variables.

Objective a   Solve systems of three equations in three variables.

Solve.

1. $x - y - 2z = 1,$
      $x - 5y + 2z = 5,$
    $2x - 3y - 4z = 2$

1. ______________

2. $a + b - c = 4,$
   $2a - b - 3c = 1,$
   $a + 2b + 3c = -1$

2. ______________

3. $x - 2y - z = 4,$
   $2x + 3y + 2z = 11,$
   $2x - y + 4z = 5$

3. ______________

4. $4x - y - z = -1,$
   $2x + 3y + z = 2,$
   $2x - y - z = -2$

4. ______________

237

5.
$$p - q + 3r = 2,$$
$$2p - 8q + 3r = 6,$$
$$10p + 4q - 9r = 0$$

5. _______________

6.
$$2a + c = 5,$$
$$5a - 2b = 0,$$
$$b + 2c = -5$$

6. _______________

7.
$$2x + y \quad\;\; = 1,$$
$$4x + 3y + z = 7,$$
$$2y + z = 2$$

7. _______________

8.
$$x + 2y - 3z = 8,$$
$$2x + y - 2z = 11,$$
$$x + 5y - 8z = 15$$

8. _______________

238

Chapter 8 SYSTEMS OF EQUATIONS

8.6 Solving Applied Problems: Three Equations

Learning Objective
a   Solve applied problems using systems of three equations.

Objective a   Solve applied problems using systems of three equations.

Solve.

1. The sum of three numbers is 78. The third is 4 more than   1. _______________
    the second. The first is 2 less than twice the second. Find
    the numbers.

2. In triangle *ABC*, the measure of angle *A* is twice the   2. _______________
    measure of angle *B*. The measure of angle *C* is 4° less than
    that of angle *B*. Find the angle measures.

239

3. The number of calories in a 100-g serving of fruit varies by the type of fruit. A snack consisting of one serving each of grapes, melon, and pears contains 124 calories. A snack consisting of two servings of grapes and one serving of pears contains 160 calories. A snack consisting of 2 servings of melon, one serving of grapes, and two servings of pears contains 188 calories. How many calories are in a 100-g serving of each type of fruit?

3. _______________

4. Surresh is going to buy a computer. The basic model with a memory upgrade and a larger monitor costs $890. The basic model with only a memory upgrade costs $840. The price of the basic model with a larger monitor is $800. Find the price of the basic model, the price of a memory upgrade, and the added cost of a larger monitor.

4. _______________

240

5. Polar Treats sells one-dip ice cream cones for $2.25, two-dip cones for $3.15, and three-dip cones for $3.85. On Saturday, they sold 200 cones for a total of $599.20. Altogether, Polar Treats used 380 dips of ice cream. How many of each size cone were sold?

5. ______________________

6. Ragheda has $11,200 invested in three retirement accounts. Last year, the first grew by 8%, the second by 10%, and the third by 5%. Her total earnings were $848. The earnings from the third investment were $80 less than the earnings from the second investment. How much was invested in each account?

6. ______________________

7. Willoughby County spent $7.6 million last year for
schools, roads, and law enforcement. The amount spent on
roads was $0.4 million more than the amount spent on law
enforcement. The amount spent on roads was $0.2 million
less than one fourth of what was spent on schools. How
much did the county spend on schools, roads, and law
enforcement?

7. ________________

8. In their final home game, the Wildcats scored a total of
113 points on a combination of 2-point field goals, 3-point
field goals, and 1-point foul shots. Altogether, the
Wildcats made 59 baskets and 22 more 2-pointers than
foul shots. How many shots of each kind were made?

8. ________________

242

Chapter 9 MORE ON INEQUALITIES

9.1 Sets, Inequalities, and Interval Notation

Learning Objectives

a   Determine whether a given number is a solution of an inequality.

b   Write interval notation for the solution set or the graph of an inequality.

c   Solve an inequality using the addition principle and the multiplication principle and then graph the inequality.

d   Solve applied problems by translating to inequalities.

Key Terms

Use the vow3cabulary terms listed below to complete each statement in Exercises 1–4.

     equivalent   **graph**   **inequality**   **set-builder notation**

1.   A(n) ___________________ is a number sentence with $<$, $>$, $\leq$, or $\geq$ as its verb.

2.   A(n) ___________________ of an inequality is a drawing that represents its solutions.

3.   The sentences $x + 4 < 10$ and $x < 6$ are ___________________ since they have the same solution set.

4.   The solution set $\{x \mid x > 2\}$ is written using ___________________ .

Objective a   Determine whether a given number is a solution of an inequality.

Determine whether each number is a solution of the given inequality.

5. $x \leq -6$ **5.**

    a) 0   a) _____________

    b) −3   b) _____________

    c) −6   c) _____________

    d) −9   d) _____________

    e) −5.4   e) _____________

6. $x > 10$

   a) 0

   b) -12

   c) 18

   d) 12.7

   e) 10

6.

   a) ______________

   b) ______________

   c) ______________

   d) ______________

   e) ______________

Objective b   Write interval notation for the solution set or the graph of an inequality.

Write interval notation for the given set.

7. $\{x \mid -3 < x < 6\}$

7. ______________

8. $x \geq 3$

8. ______________

9. $x < -4$

9. ______________

Objective c   Solve an inequality using the addition principle and the multiplication principle and then graph the inequality.

Solve and graph.

10. $x + 6 > 3$

10. ______________

____

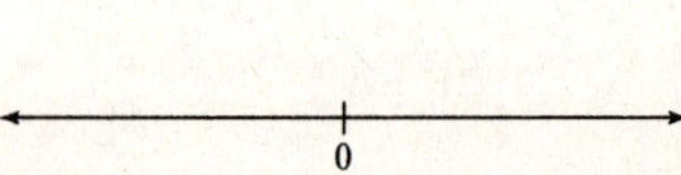

244

11. $2x + 4 \leq x + 7$

11. ____________

____

12. $-12x > -36$

12. ____________

____

13. $3x \geq 18$

13. ____________

____

Solve.

14. $y + \dfrac{2}{7} \leq \dfrac{7}{14}$

14. ____________

____

15. $-10z + 5 > 5 - 11z$

15. ____________

____

16. $9x \geq -8$

16. ____________

____

245

17. $\dfrac{-4}{7} > -6x$

17.____________

____

18. $6 + 9n < -21$

18.____________

____

19. $8x - 8 \leq 32$

19.____________

____

20. $9x - 9 < -54$

20.____________

____

21. $5x + 2 - 4x \leq 13$

21.____________

____

22. $6 - 20x \leq 5 - 11x - 8x$

22.____________

____

23. $\dfrac{x}{4} - 1 \leq \dfrac{1}{2}$

23.____________

____

246

24. $5(2x-3) < 45$ 24.______________

                                                          ____

25. $5(2y-9) \geq 6(3y+4)$ 25.______________

                                                          ____

Objective d   Solve applied problems by translating to inequalities.

Solve.

26. You are taking a math course in which there will   26.______________
be four tests, each worth 100 points. You have
scores of 97, 94, and 97 on the first three tests.
You must earn a total of 360 points in order to get
an A. What scores on the last test will give you an
A?

27. In planning for a banquet, you find that one   27.______________
speaker charges $225 plus 50% of the total ticket
sales. Another speaker charges a flat fee of $525.
In order for the first speaker to produce more
profit for your organization than the other
speaker, what is the highest price you can charge
per ticket, assuming that 200 people will attend?

28. Most insurance companies will replace a vehicle
any time an estimated repair exceeds 80% of its
"blue book" value. Melissa's car had $6500 in
repairs after an accident. What can be concluded
about its "blue book" value?

28.______________

29. Carlos can be paid in one of two ways. Plan A is
a salary of $500 per month, plus a commission of
8% of sales. Plan B is a salary of $653 per
month, plus a commission of 5% of sales. For
what amount of sales is Carlos better off
selecting Plan A?

29.______________

248

Chapter 9 MORE ON INEQUALITIES

9.2 Intersections, Unions, and Compound Inequalities

Learning Objective

a   Find the intersection of two sets. Solve and graph conjunctions of inequalities.
b   Find the union of two sets. Solve and graph disjunctions of inequalities.
c   Solve applied problems involving conjunctions and disjunctions of inequalities.

Key Terms
Use the vocabulary terms listed below to complete each statement in Exercises 1–5.

conjunction	**compound inequalities**	**disjunction**
union	**intersection**	

1. Two inequalities joined by the word "and" or the word "or" are called

   __________________.

2. A(n) __________________ is a compound sentence formed using the word "and."

3. The solution of a conjunction is the __________________ of the solutions sets of the individual sentences.

4. A(n) __________________ is a compound sentence formed using the word "or."

5. The solution of a disjunction is the __________________ of the solutions sets of the individual sentences.

Find the intersection or union.

6. $\{5,6,8,9\} \cap \{4,6,9,11\}$ 6._______________

7. $\{7,8,9,10,11\} \cup \{5,7,9,13\}$ 7._______________

8. $\{3,8,20,24\} \cup \{2,3,8,15\}$ 8._______________

9. $\{c,d,e,f,g\} \cap \{f,g,h,i,j\}$ 9._______________

249

10. $\{m,a,t,h\} \cap \{m,a,t\}$

10._____________

11. $\{2,3,5,10\} \cup \varnothing$

11._____________

Objective a **Find the intersection of two sets. Solve and graph conjunctions of inequalities.**

Graph and write interval notation.

12. $-2 \leq x \leq 5$

12._____________

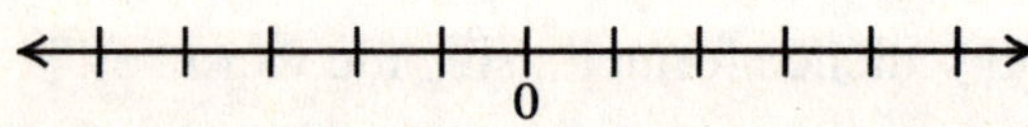

13. $6 \geq m$ *and* $m > -2$

13._____________

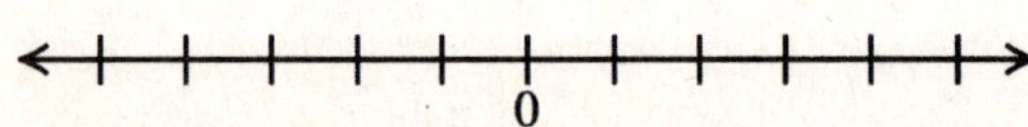

14. $x < 5$ *and* $x \geq 2$

14._____________

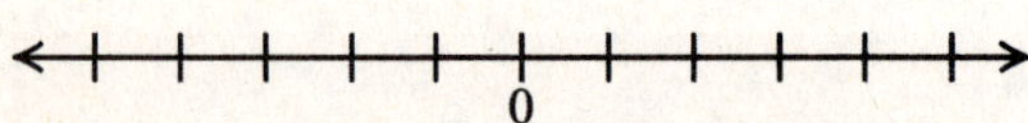

Solve and graph.

15. $-4 \leq x - 5 < 3$

15._____________

16. $-2 < 3t + 1$ *and* $2t - 5 \leq 7$

16._____________

250

Solve.

17. $5 \geq \dfrac{x+1}{4} \geq -3$ 17.___________

18. $-4 < 3x + 2 \leq 11$ 18.___________

Objective b   Find the union of two sets. Solve and graph disjunctions of inequalities.

Graph and write interval notation.

19. $t < 0 \ \ or \ \ t > 5$ 19.___________

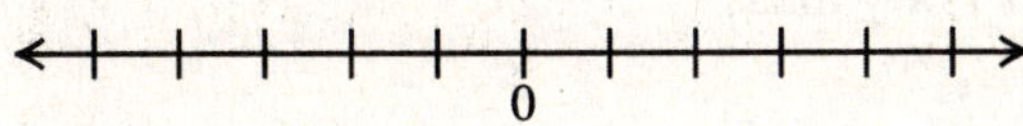

20. $y \geq -2 \ \ or \ \ -y \geq 3$ 20.___________

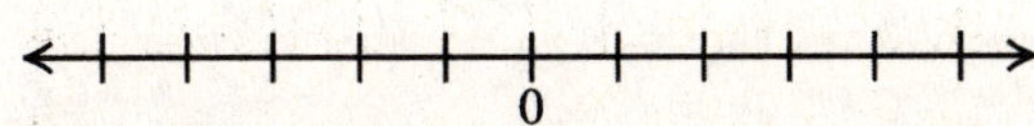

21. $x \geq -3 \ \ or \ \ x \geq -1$ 21.___________

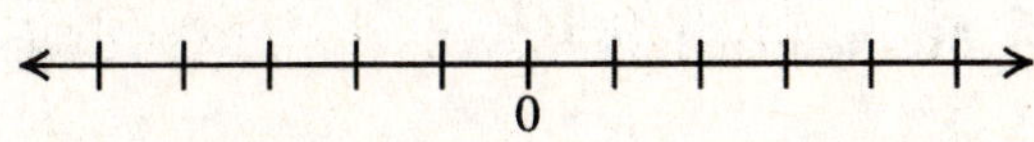

Solve and graph.

22. $2z + 3 \geq -5 \ \ or \ \ 6z - 7 < 11$ 22.___________

23. $3a - 5 \geq 8 \ \ or \ \ 7 - 2a < 1$ 23.___________

251

Solve.

24. $5x+2<-3$ or $5x-2>3$

24.______________

25. $-3m+4<10$ or $-3m-4\geq5$

25.______________

Objective c   Solve applied problems involving conjunctions and disjunctions of inequalities.

Solve.

26. Jen's height is 65 in. Use the formula $I=\dfrac{703w}{H^2}$ to find the weight w that will allow Jen to keep her body mass index I within the 18.5-24.9 range.

26.______________

27. Jack's height is 74 in. Use the formula $I=\dfrac{703w}{H^2}$ to find the weight w that will allow Jack to keep his body mass index I within the 18.5-24.9 range.

27.______________

28. The dosage of a medication for a 10-year old child must stay between 150 mg and 250 mg. Use the formula $c=\dfrac{ac}{a+12}$, where a is the child's age, and d is the usual adult dosage, to find the equivalent adult dosage.

28.______________

252

Chapter 9 MORE ON INEQUALITIES

9.3 Absolute-Value Equations and Inequalities

Learning Objective
a   Simplify expressions containing absolute-value symbols.
b   Find the distance between two points on the number line.
c   Solve equations with absolute-value expressions.
d   Solve equations with two absolute-value expressions.
e   Solve inequalities with absolute-value expressions.

Key Terms
Finish each statement in Exercises 1–5 with p, $-p$, or 0, where p is a positive number.

1.   The solutions of $|X| = $ _________ are those numbers that satisfy $X = p$ or $X = -p$.

2.   The equation $|X| = $ _________ is equivalent to the equation $X = 0$.

3.   The equation $|X| = $ _________ has no solution.

4.   The solutions of $|X| < p$ are those numbers that satisfy _________ $< X < $ _________.

5.   The solutions of $|X| > p$ are those numbers that satisfy $X < $ _________ or
   $X > $ _________.

Objective a   Simplify expressions containing absolute-value symbols.

Simplify, leaving as little as possible inside absolute-value signs.

6. $|7x|$ 6.__________________

7. $|5x^2|$ 7.__________________

8. $|-6x^2|$ 8.__________________

9. $|-19z|$

9._____________

10. $\left|\dfrac{y^2}{-z}\right|$

10._____________

11. $\left|\dfrac{-8a^2}{4a}\right|$

11._____________

12. $\left|\dfrac{5x^2}{-15x}\right|$

12._____________

13. $\left|\dfrac{72x^5}{-8x}\right|$

13._____________

Objective b   Find the distance between two points on the number line.

Find the distance between the points on the number line.

14. $-11, -96$

14._____________

15. $-56, 27$

15._____________

16. $-2.6, -4.8$

16._____________

17. $\dfrac{3}{4}, -\dfrac{7}{12}$

17._____________

254

Objective c   Solve equations with absolute-value expressions.

Solve.

18. $|x| = 4.3$ 18.___________________

19. $|t| = 0$ 19.___________________

20. $|x| = -18$ 20.___________________

21. $|4x - 5| = 8$ 21.___________________

22. $|x - 7| = 10$ 22.___________________

23. $|y - 5| = -15$ 23.___________________

24. $|3x - 7| = 4$ 24.___________________

25. $|5x| - 7 = 9$

25.___________________

26. $\left|\dfrac{3-5t}{2}\right| = 11$

26.___________________

Objective d   Solve equations with two absolute-value expressions.

Solve.

27. $|2x + 1| = |x - 6|$

27.___________________

28. $|x + 6| = |x - 7|$

28.___________________

29. $|5t + 4| = |2t - 7|$

29.___________________

256

30. $\left|3x-4\right|=\left|4-3x\right|$

30._______________

Objective e   Solve inequalities with absolute-value expressions.

Solve.

31. $\left|x\right|\geq 3$

31._______________

32. $\left|x\right|<15$

32._______________

33. $\left|x+3\right|<7$

33._______________

34. $\left|x-3\right|+8\geq 14$

34._______________

35. $13-\left|2x+1\right|\leq 5$

35._______________

36. $\left|6-3a\right| \le 4$

36._______________

37. $\left|\dfrac{7-2x}{3}\right| \ge \dfrac{5}{2}$

37._______________

Chapter 9 MORE ON INEQUALITIES

9.4 Systems of Inequalities in Two Variables

Learning Objectives

a   Determine whether an ordered pair of numbers is a solution of an inequality in two variables.

b   Graph linear inequalities in two variables.

c   Graph systems of linear inequalities and find coordinates of any vertices.

Key Terms

Use the vocabulary terms listed below to complete each statement in Exercises 1–4.

half-plane	**linear inequality**	**test point**
vertices		

1.   The sentence $2x - 3 \leq -5$ is an example of a(n) _______________________ .

2.   The graph of a linear inequality is a(n) _______________________ .

3.   To determine which side of the boundary to shade as the graph of the solution set of an inequality, select a(n) _______________________ not on the line.

4.   If a system of inequalities has a graph that consists of a polygon and its interior, the corners of the graph are called _______________________ .

Objective a   Determine whether an ordered pair of numbers is a solution of an inequality in two variables.

Determine whether the ordered pair is a solution of the given inequality.

5. $(2, -5)$; $2x + 4y \leq -5$ 5. _______________

6. $(20, 12)$; $2y - 3x > 1$ 6. _______________

Objective b   Graph linear inequalities in two variables.

Graph each inequality on a plane.

7. $y \le \frac{1}{3}x$

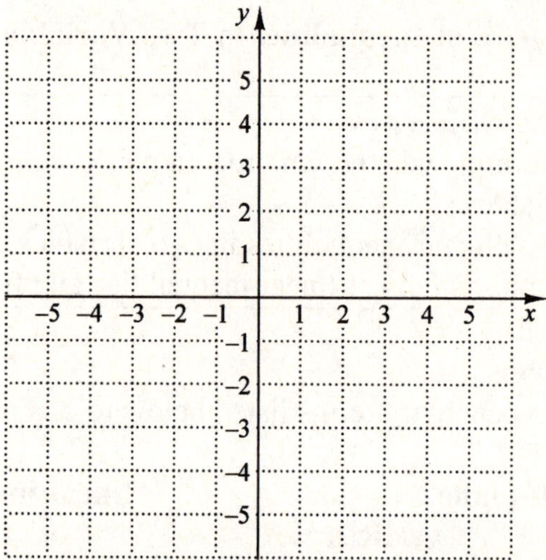

8. $y \ge x - 2$

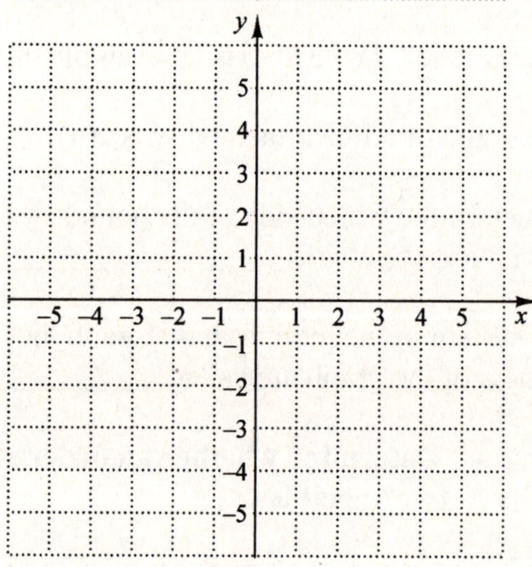

9. $4x - y > 8$

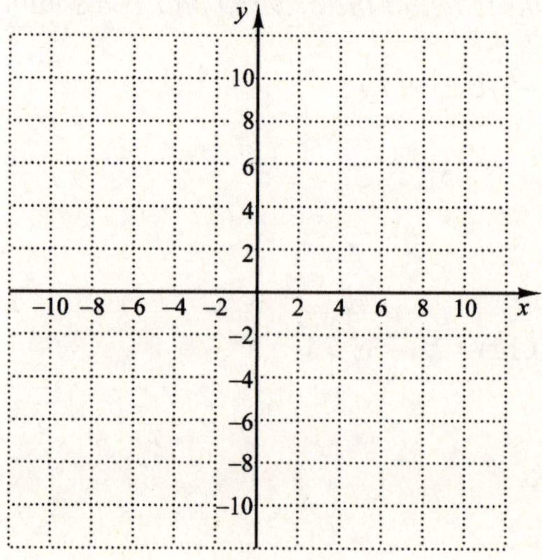

260

10. $x - 3y < 6 - x$

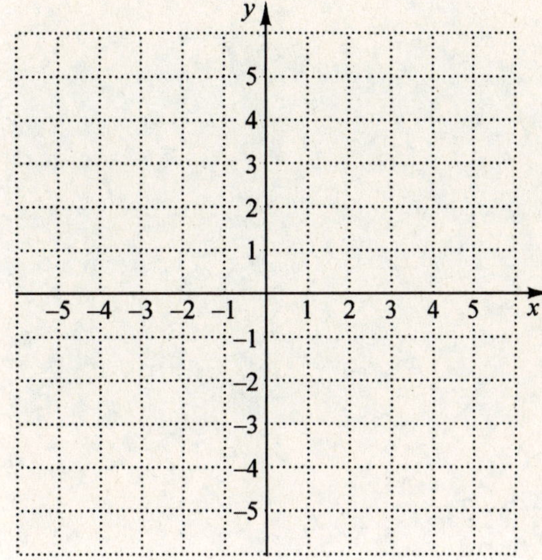

Objective c   Graph systems of linear inequalities and find coordinates of any vertices.

Graph each system of inequalities.

11. $y < -x$,
         $y > x - 3$

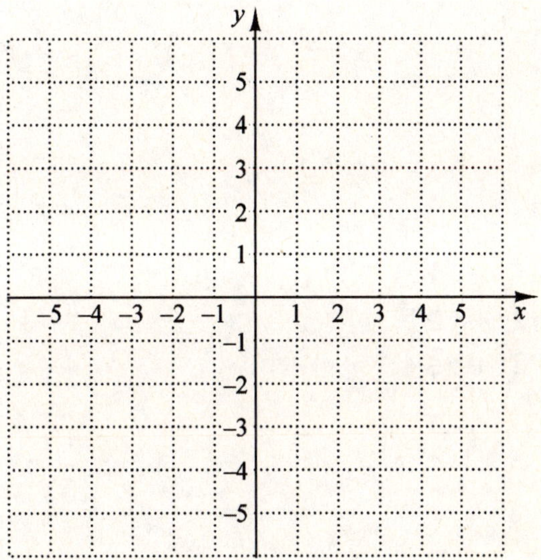

12. $y \le 2x$,
$\quad y \ge 4x - 3$

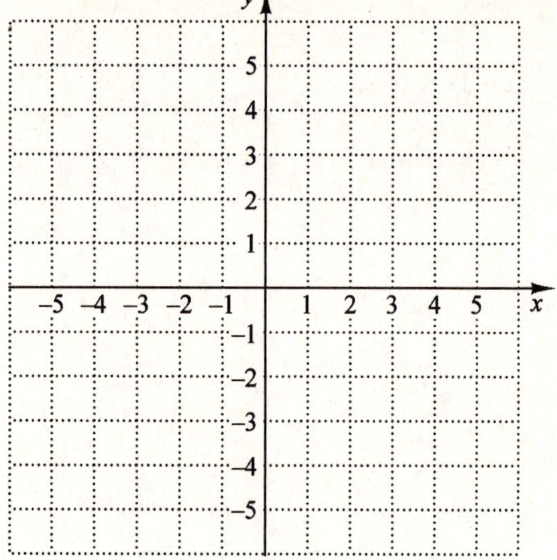

13. $y \ge -3$,
$\quad y \le 2 - x$

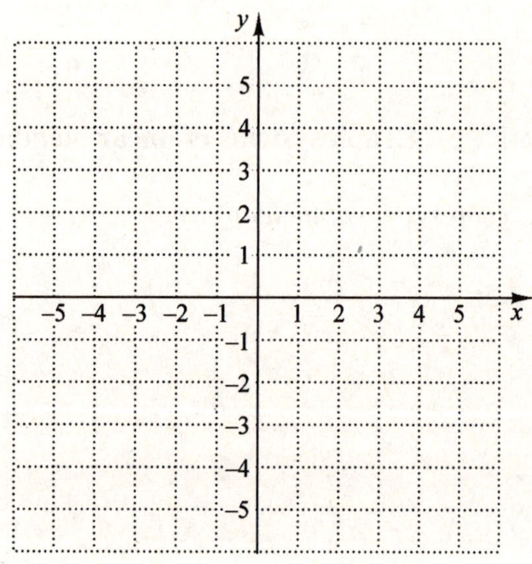

14. $2x + y \ge 1$,
$\quad 2x + y \le 4$

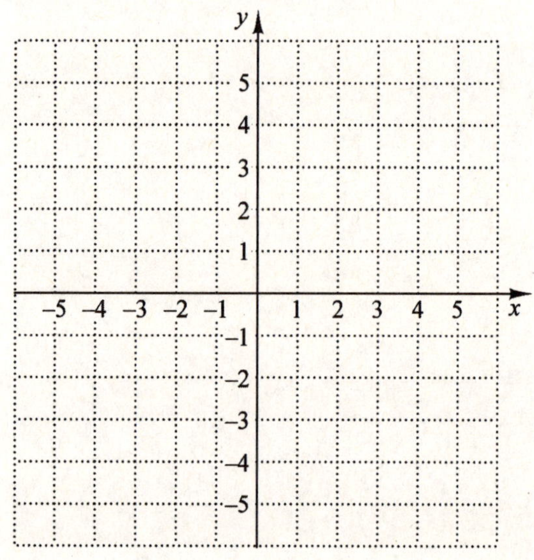

262

15. $y \geq x - 6$,
$\quad\ y \leq 3x - 4$,
$\quad\ x \leq 5$

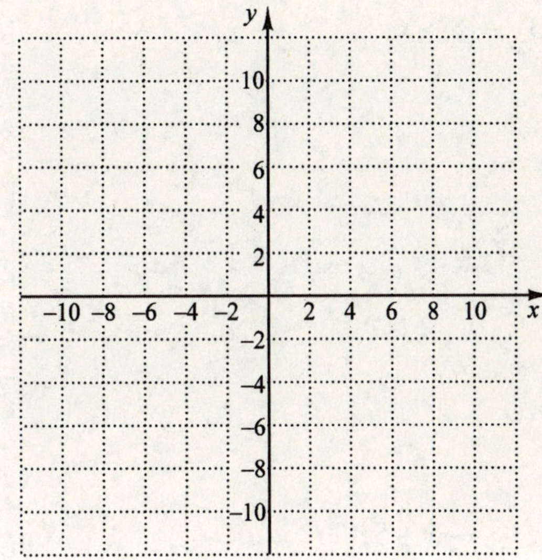

16. $2x + 4y \leq 24$,
$\quad\ 5x + 3y \leq 30$,
$\quad\ x \geq 0$,
$\quad\ y \geq 0$

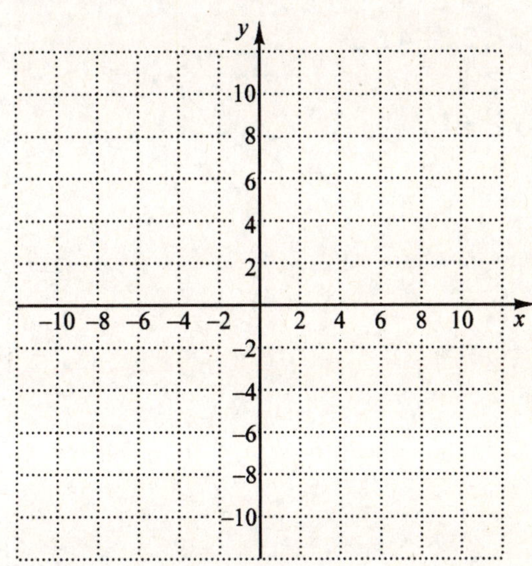

17. $-4 > x - 2y$,
$\quad\ x + 6y < 36$,
$\quad\ 2y \geq -3x - 12$

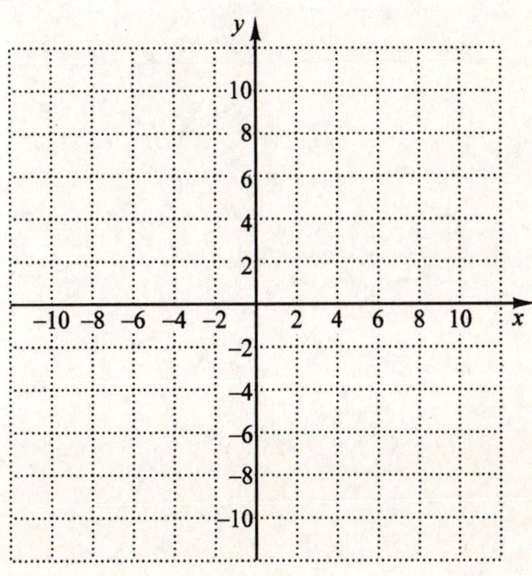

263

18. $y - 4 > 2x$,
$y > -6x$,
$3y - 18 > x$

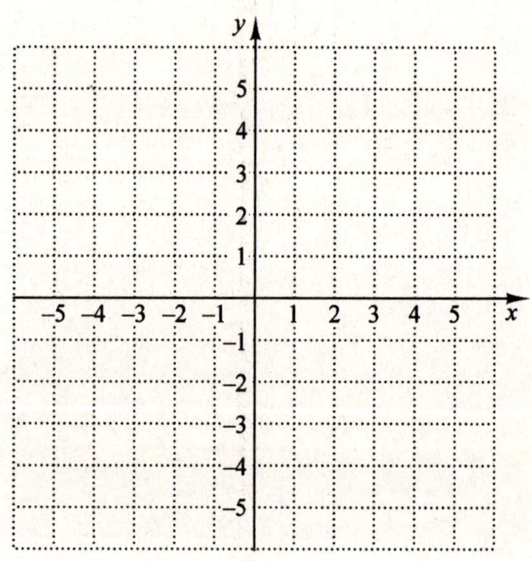

19. $y \leq -\dfrac{1}{2}x + 5$,
$y < 2x + 4$,
$x < 4$,
$y \geq -1$

Chapter 10 RADICAL EXPRESSIONS, EQUATIONS, AND FUNCTIONS

10.1 Radical Expressions and Functions

Learning Objectives

a   Find principal square roots and their opposites, approximate square roots, identify radicands, find outputs of square-root functions, graph square-root functions, and find the domains of square-root functions.

b   Simplify radical expressions with perfect-square radicands.

c   Find cube roots, simplifying certain expressions, and find outputs of cube-root functions.

d   Simplify expressions involving odd roots and even roots.

Key Terms

Use the vocabulary terms listed below to complete each statement in Exercises 1–6.

cube	**index**	**irrational**
principal	**radicand**	**square**

1.   The number c is a(n) _________________ root of a if $c^2 = a$.

2.   The _________________ square root of a nonnegative number is its nonnegative square root.

3.   The square root of any whole number that is not a perfect square is a(n) _________________ number.

4.   The _________________ is the expression under the radical sign.

5.   The number c is a(n) _________________ root of a if $c^3 = a$.

6.   The number n in the expression $\sqrt[n]{a}$ is called the _________________.

Objective a   Find principal square roots and their opposites, approximate square roots, identify radicands, find outputs of square-root functions, graph square-root functions, and find the domains of square-root functions.

Find the square roots.

7.   64   7. _________________

265

8. 4900

Simplify.

9. $-\sqrt{\dfrac{225}{169}}$

10. $\sqrt{256}$

11. $\sqrt{0.09}$

12. $-\sqrt{0.0121}$

Use a calculator to approximate to three decimal places.

13. $\sqrt{562}$

14. $\sqrt{\dfrac{893}{147}}$

15. $-\sqrt{762.8}$

8. _______________

9. _______________

10. _______________

11. _______________

12. _______________

13. _______________

14. _______________

15. _______________

266

Name: Date:
Instructor: Section:

Identify the radicand.

16. $7\sqrt{x+1}+5$

16. _______________

17. $s\sqrt[3]{\dfrac{2s}{3t}}$

17. _______________

For the given function, find the indicated function values.

18. $f(x)=\sqrt{3x-12}$; $f(5), f(4), f(0), f(-5)$

18. _______________

19. $g(x)=\sqrt{10-x^2}$; $g(5), g(3), g(0), g(-2)$

19. _______________

20. Find the domain of the function $g(x)=\sqrt{x-4}$.

20. _______________

21. $f(x) = \sqrt{x+5}$

21.

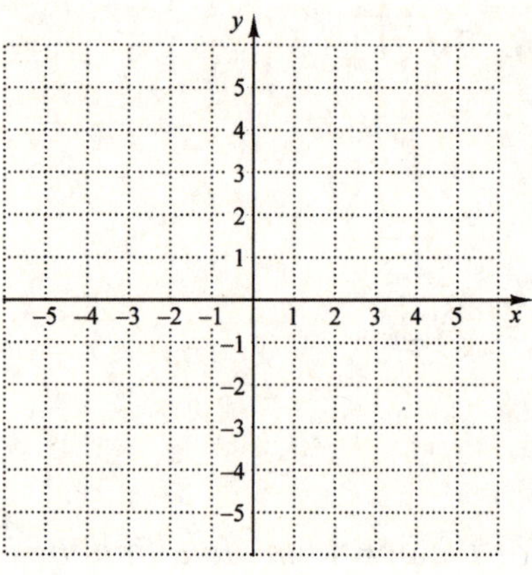

22. $g(x) = \sqrt{x} + 5$

22.

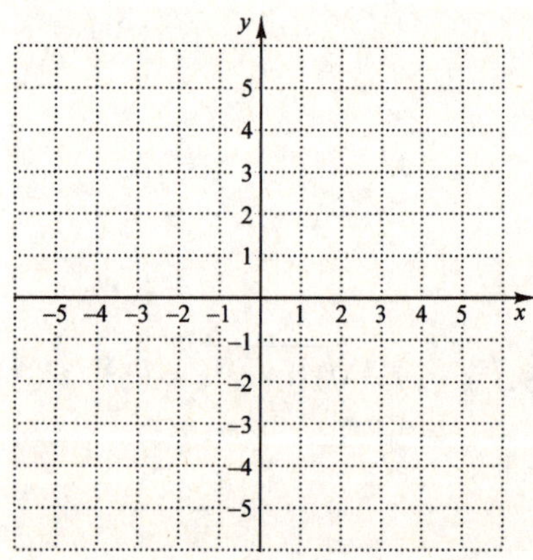

23. $g(x) = \sqrt{x} - 2$

23.

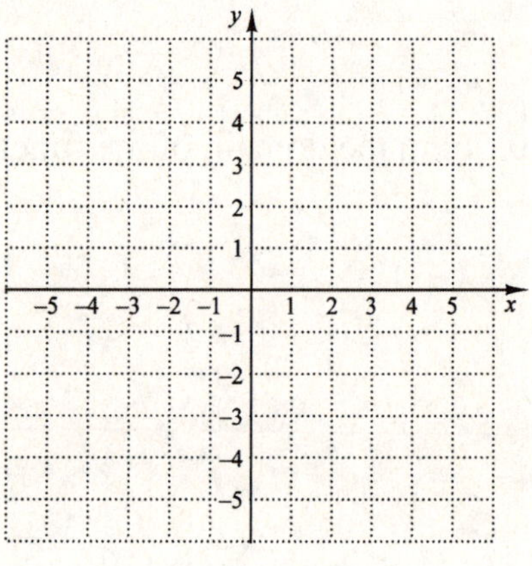
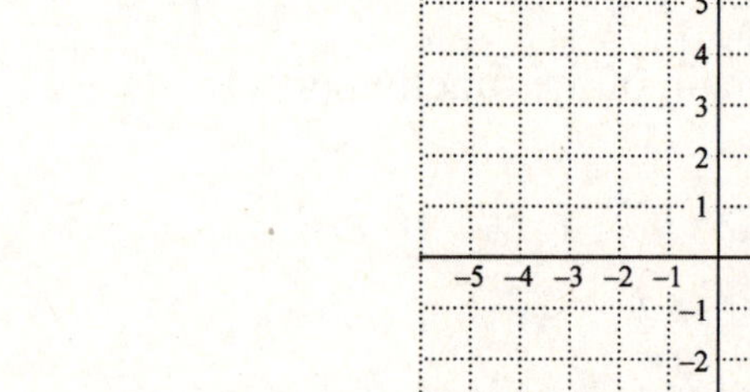

Objective b   Simplify radical expressions with perfect-square radicands.

Find each of the following. Assume that letters can represent any *real number.*

24. $\sqrt{100m^2}$

24. ________________

25. $\sqrt{y^2 - 10y + 25}$

25. ________________

26. $\sqrt{81c^2}$

26. ________________

27. $\sqrt{4x^2 - 28x + 49}$

27. ________________

Objective c   Find cube roots, simplifying certain expressions, and find outputs of cube-root functions.

Simplify.

28. $\sqrt[3]{-8}$

28. ________________

29. $\sqrt[3]{-125t^3}$

29. ________________

269

For the given function, find the indicated function values.

30. $g(x) = \sqrt[3]{2x-1}$; $g(0)$, $g(14)$, $g(-13)$, $g(-62)$

30. _________________

31. $f(t) = \sqrt[4]{t-10}$; $f(9)$, $f(11)$, $f(-6)$, $f(26)$

31. _________________

Objective d   Simplify expressions involving odd roots and even roots.

Find each of the following. Assume that letters can represent any *real number.*

32. $\sqrt[6]{(5x)^6}$

32. _________________

33. $\sqrt[3]{-\dfrac{8}{27}}$

33. _________________

34. $\sqrt[4]{10,000}$

34. _________________

35. $\sqrt[9]{(3ab)^9}$

35. _________________

270

Chapter 10 RADICAL EXPRESSIONS, EQUATIONS, AND FUNCTIONS

10.2 Rational Numbers as Exponents

Learning Objective
a   Write expressions with or without rational exponents, and simplify, if possible.
b   Write expressions without negative exponents, and simplify, if possible.
c   Use the laws of exponents with rational exponents.
d   Use rational exponents to simplify radical expressions.

Key Terms
In Exercises 1–6, match each expression with an equivalent expression in the right column.

1. _____ a^{-n} **a)** a^{m-n}

2. _____ $a^{m/n}$ **b)** $\sqrt[n]{a^m}$

3. _____ $a^m \cdot a^n$

                                         c) $\dfrac{1}{a^n}$

4. _____ $\dfrac{a^m}{a^n}$

                                         d) a^{m+n}

5. _____ $(a^m)^n$ **e)** $a^m b^m$

6. _____ $(ab)^m$ **f)** a^{mn}

Objective a   Write expressions with or without rational exponents, and simplify, if possible.

Rewrite without rational exponents, and simplify, if possible.

7. $a^{1/4}$ 7. ________________

8. $1000^{1/3}$ 8. ________________

9. $\left(pqr\right)^{1/2}$

9. _______________

10. $\left(x^3 y^4\right)^{1/5}$

10. _______________

11. $8^{5/3}$

11. _______________

12. $\left(36c^6\right)^{3/2}$

12. _______________

Rewrite without rational exponents.

13. $\sqrt[5]{10}$

13. _______________

14. $\sqrt{x^7}$

14. _______________

15. $\sqrt[4]{ab^3}$

15. _______________

16. $\left(\sqrt[5]{3mn^4}\right)^2$

16. _______________

272

Objective b   Write expressions without negative exponents, and simplify, if possible.

Rewrite with positive exponents, and simplify, if possible.

17. $(3x)^{-1/5}$

17. _______________

18. $\left(\dfrac{1}{27}\right)^{-2/3}$

18. _______________

19. $\dfrac{3x^2}{y^{-4/5}}$

19. _______________

20. $9^{-1/2}a^{2/3}b^{-3/5}$

20. _______________

Objective c   Use the laws of exponents with rational exponents.

Use the laws of exponents to simplify. Write the answers with positive exponents.

21. $3^{2/5}\cdot 3^{1/10}$

21. _______________

22. $\dfrac{5^{-1/3}}{5^{-3/4}}$

22. _______________

23. $\left(x^{4/5}\right)^{3/2}$

23. _______________

24. $\left(a^{2/5}b^{-1/5}\right)^{-1/2}$

24. _______________

273

Objective d   Use rational exponents to simplify radical expressions.

Use rational exponents to simplify. Write the answer in radical notation if appropriate.

25. $\sqrt[3]{x^{15}}$

25. _______________

26. $\left(\sqrt[6]{cd}\right)^3$

26. _______________

27. $\sqrt{\sqrt[5]{m}}$

27. _______________

28. $\left(\sqrt[4]{x^3 y^2}\right)^{12}$

28. _______________

Use rational exponents to write a single radical expression.

29. $\sqrt[3]{6}\sqrt{6}$

29. _______________

30. $\sqrt[5]{\sqrt[3]{x}}$

30. _______________

31. $\left(\sqrt[4]{a^3 b^2}\right)^{12}$

31. _______________

32. $\dfrac{x^{7/12} \cdot y^{5/12}}{x^{1/3} \cdot y^{-1/2}}$

32. _______________

Chapter 10 RADICAL EXPRESSIONS, EQUATIONS, AND FUNCTIONS

10.3 Simplifying Radical Expressions

Learning Objectives
a   Multiply and simplify radical expressions.
b   Divide and simplify radical expressions.

Key Terms
Use the vocabulary terms listed below to complete each statement in Exercises 1–4.

index   **multiples**   **radicands**   **square**

1. The product of two nth roots is the nth root of the product of the two

    _______________________.

2. The product rule for radicals applies only when radicals have the same

    _______________________ .

3. When no index is written, roots are understood to be _____________________ roots.

4. To simplify an nth root, identify factors in the radicand with exponents that are

    _______________________ of n.

Objective a   Multiply and simplify radical expressions.

Simplify by factoring. Assume that no radicands were formed by raising negative numbers to even powers.

5. $\sqrt{200}$ 5._______________

6. $\sqrt{63}$ 6._______________

7. $\sqrt[3]{27x^6y^2}$

7.___________________

8. $-\sqrt[3]{24x^{12}}$

8.___________________

9. $\sqrt{x^5y^{10}}$

9.___________________

10. $\sqrt[3]{a^6b^7c^{14}}$

10.___________________

11. $\sqrt[5]{-320c^8d^{17}}$

11.___________________

12. $\sqrt[4]{32x^3y^{20}z^{11}}$

12.___________________

Multiply and simplify. Assume that no radicands were formed by raising negative numbers to even powers.

13. $\sqrt{10}\sqrt{11}$

13.___________________

14. $\sqrt[5]{3x^2}\sqrt[5]{5x}$

14.___________________

15. $\sqrt{14}\sqrt{21}$

15.___________________

16. $\sqrt[3]{5}\sqrt[3]{25}$

16.___________________

17. $\sqrt{28x^5}\sqrt{28x^5}$

17.___________________

276

18. $\sqrt[3]{4a^2}\,\sqrt[3]{6a^5}$ 18._____________

19. $\sqrt[3]{c^4d^8}\,\sqrt[3]{cd^2}$ 19._____________

20. $\sqrt[4]{50x^5y^3}\,\sqrt[4]{75x^2y}$ 20._____________

Objective b   Divide and simplify radical expressions.

Divide and simplify. Assume that all expressions under radicals represent positive numbers.

21. $\dfrac{\sqrt{30x}}{\sqrt{5x}}$ 21._____________

22. $\dfrac{\sqrt[3]{30}}{\sqrt[3]{3}}$ 22._____________

23. $\dfrac{\sqrt{6a^5b}}{\sqrt{2a}}$ 23._____________

24. $\dfrac{\sqrt{81xyz}}{3\sqrt{3}}$ 24._____________

Simplify.

25. $\sqrt{\dfrac{81}{49}}$

25. _______________

26. $\sqrt[3]{\dfrac{125}{64}}$

26. _______________

27. $\sqrt{\dfrac{16p^5}{q^{10}}}$

27. _______________

28. $\sqrt[3]{\dfrac{1000x^7}{27y^3}}$

28. _______________

29. $\sqrt[4]{\dfrac{a^8b^7}{c^{14}}}$

29. _______________

30. $\sqrt[6]{\dfrac{m^7n^6}{p^{20}}}$

30. _______________

278

Chapter 10 RADICAL EXPRESSIONS, EQUATIONS, AND FUNCTIONS

10.4 Addition, Subtraction, and More Multiplication

> **Learning Objectives**
> a   Add or subtract with radical notation and simplify.
> b   Multiply expressions involving radicals in which some factors contain more than one term.

Key Terms
Use the vocabulary terms listed below to complete each statement in Exercises 1–2.

    conjugates   **like radicals**

1. When two radical expressions have the same indices and radicands, they are called

    __________________ .

2. Pairs of radical terms like $\sqrt{a} + \sqrt{b}$ and $\sqrt{a} - \sqrt{b}$ are called __________________ .

Objective a   Add or subtract with radical notation and simplify.

Add or subtract. Then simplify by collecting like radical terms, if possible. Assume that no radicands were formed by raising negative numbers to even powers.

3. $5\sqrt{11} - 8\sqrt{11}$ 3. __________________

4. $\sqrt[5]{x} + 15\sqrt[5]{x}$ 4. __________________

5. $4\sqrt{3} + 2\sqrt[3]{5} - \sqrt{3} + 8\sqrt[3]{5}$ 5. __________________

6. $6\sqrt{18} - 5\sqrt{8}$ 6. __________________

7. $\sqrt[3]{40a^7} + 9\sqrt[3]{5a^4}$

7. _______________

8. $3\sqrt{98} - 5\sqrt{50}$

8. _______________

9. $\sqrt{16z - 4} - \sqrt{4z^3 - z^2}$

9. _______________

Objective b Multiply expressions involving radicals in which some factors contain more than one term.

Multiply. Assume that no radicands were formed by raising negative numbers to even powers.

10. $2\sqrt{3}\left(\sqrt{3} - \sqrt{10}\right)$

10. _______________

11. $\sqrt[3]{4}\left(3\sqrt[3]{2} + 5\sqrt[3]{10}\right)$

11. _______________

12. $\left(3 + \sqrt{5}\right)\left(7 - \sqrt{5}\right)$

12. _______________

13. $\left(2\sqrt{5} + 3\sqrt{2}\right)\left(3\sqrt{5} - 6\sqrt{2}\right)$

13. _______________

14. $\left(6 + \sqrt{3}\right)\left(6 - \sqrt{3}\right)$

14. _______________

15. $\left(\sqrt{2x} - \sqrt{7}\right)^2$

15. _______________

16. $\left(\sqrt[3]{9} - \sqrt[3]{3}\right)\left(\sqrt[3]{2} + \sqrt[3]{5}\right)$

16. _______________

17. $\left(\sqrt[4]{9} - \sqrt[4]{3}\right)\left(\sqrt[4]{8} + \sqrt[4]{25}\right)$

17. _______________

280

Chapter 10 RADICAL EXPRESSIONS, EQUATIONS, AND FUNCTIONS

10.5 More on Division of Radical Expressions

Learning Objective
a   Rationalize the denominator of a radical expression having one term in the denominator. b   Rationalize the denominator of a radical expression having two terms in the denominator.

Objective a   Rationalize the denominator of a radical expression having one term in the denominator.

Rationalize the denominator. Assume that no radicands were formed by raising negative numbers to even powers.

1. $\dfrac{3\sqrt{7}}{5\sqrt{2}}$

1. _______________

2. $\dfrac{\sqrt[3]{2s}}{\sqrt[3]{3t}}$

2. _______________

3. $\sqrt[3]{\dfrac{10}{xy^2}}$

3. _______________

4. $\sqrt{\dfrac{8}{50x^2y}}$

4. _______________

5. $\sqrt{\dfrac{15}{35}}$

5. _______________

6. $\dfrac{5\sqrt{11}}{4\sqrt{3}}$

6. _______________

7. $\dfrac{\sqrt[3]{5}}{\sqrt[3]{7}}$

7. _______________

8. $\sqrt[3]{\dfrac{6x^7}{5y}}$

8. _______________

Objective b   Rationalize the denominator of a radical expression having two terms in the denominator.

Rationalize the denominator. Assume that no radicands were formed by raising negative numbers to even powers.

9. $\dfrac{2}{5+\sqrt{3}}$

9. _______________

10. $\dfrac{6+\sqrt{7}}{2-\sqrt{5}}$

10. _______________

11. $\dfrac{\sqrt{p}}{\sqrt{p}-\sqrt{q}}$

11. _______________

12. $\dfrac{2\sqrt{11}-4\sqrt{5}}{6\sqrt{2}+3\sqrt{5}}$

12. _______________

282

Chapter 10 RADICAL EXPRESSIONS, EQUATIONS, AND FUNCTIONS

10.6 Solving Radical Equations

> **Learning Objective**
> a   Solve radical equations with one radical term.
> b   Solve radical equations with two radical terms.
> c   Solve applied problems involving radical equations.

Key Terms
Use the vocabulary terms listed below to complete each statement in Exercises 1 and 2.

principle of powers   **radical equation**

1. A _____________________ is an equation in which the variable appears in a radicand.

2. The _____________________ states that if $a = b$, then $a^n = b^n$ for any exponent n.

Objective a   Solve radical equations with one radical term.

Solve.

3. $\sqrt{3x - 7} = 5$ 3. _______________

4. $\sqrt{6x} - 3 = 7$ 4. _______________

5. $\sqrt{n - 8} - 2 = 1$ 5. _______________

6. $\sqrt{z + 2} + 6 = 10$ 6. _______________

7. $\sqrt[3]{t+3} = 4$

7. _______________

8. $\sqrt[4]{y-9} = 2$

8. _______________

9. $7\sqrt{a} = a$

9. _______________

10. $6\sqrt{x} - 10 = 2$

10. _______________

11. $\sqrt[3]{x} = -10$

11. _______________

12. $\sqrt{n+5} = -6$

12. _______________

Objective b   Solve radical equations with two radical terms.

Solve.

13. $\sqrt{6z+1} = \sqrt{5z+3}$

13. _______________

14. $7 + 2\sqrt{x+1} = x$

14. _______________

284

15. $\sqrt{5x+1} = 1 + \sqrt{4x-3}$

15. _______________

16. $7 + \sqrt{10-z} = 8 + \sqrt{1-z}$

16. _______________

17. $2\sqrt{3y-2} - \sqrt{4y-3} = 1$

17. _______________

Objective c   Solve applied problems involving radical equations.

Solve.

18. How far can you see to the horizon from a given height?
The function $D = 1.2\sqrt{h}$ can be used to approximate the
distance D, in miles, that a person can see to the horizon
from a height h, in feet.
A person can see 210 mi to the horizon from an airplane
window. How high is the airplane?

18. _______________

19. How far can you see to the horizon from a given height?
The function $D = 1.2\sqrt{h}$ can be used to approximate the
distance D, in miles, that a person can see to the horizon
from a height h, in feet.
Maddie can see 29.2 mi to the horizon from the top of a
cliff. What is the height of his eyes?

19. _______________

20. How far can you see to the horizon from a given height? The function $D = 1.2\sqrt{h}$ can be used to approximate the distance D, in miles, that a person can see to the horizon from a height h, in feet.

A technician can see 35.2 mi to the horizon from the top of a radio tower. How high is the tower?

20. __________________

21. The formula $r = 2\sqrt{5L}$ can be used to approximate the speed r, in miles per hour, of a car that has left a skid mark of length L, in feet.

How far will a car skid at 60 mph?

21. __________________

22. The formula $r = 2\sqrt{5L}$ can be used to approximate the speed r, in miles per hour, of a car that has left a skid mark of length L, in feet.

How far will a car skid at 70 mph?

22. __________________

23. The formula $T = 2\pi\sqrt{L/32}$ can be used to find the period T, in seconds, of a pendulum of length L, in feet. The pendulum in a grandfather clock has a period of 2.1 sec. Find the length of the pendulum. Use 3.14 for π.

23. __________________

24. The formula $T = 2\pi\sqrt{L/32}$ can be used to find the period T, in seconds, of a pendulum of length L, in feet. A playground swing has a period of 3.2 sec. Find the length of the swing's chain. Use 3.14 for π.

24. __________________

286

Chapter 10 RADICAL EXPRESSIONS, EQUATIONS, AND FUNCTIONS

10.7 Applications Involving Powers and Roots

Learning Objectives
a   Solve applied problems involving the Pythagorean theorem and powers and roots.

Objective a   Solve applied problems involving the Pythagorean theorem and powers and roots.

In a right triangle, find the length of a side not given. Give an exact answer and, where appropriate, an approximation to three decimal places.

1. $a = 4,\ b = 7$ 1. _______________

2. $a = 7,\ b = 11$ 2. _______________

3. $b = 4,\ c = 9$ 3. _______________

4. $c = 12,\ a = \sqrt{7}$ 4. _______________

5. $b = 2,\ c = \sqrt{26}$ 5. _______________

6. $b = 3,\ a = \sqrt{n}$ 6. _______________

In the following problems, give an exact answer and, where appropriate, an approximation to three decimal places.

7. How long is a guy wire reaching from the top of a 12-ft pole to a point on the ground 6 ft from the pole?

7. 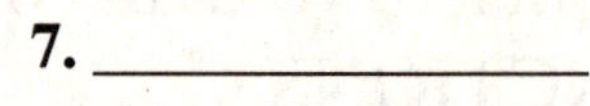

8. Using the formula $L = \dfrac{0.000169d^{2.27}}{h}$, find the length L of a road-pavement message when $h = 4$ ft and $d = 190$ ft.

8. 

9. A large screen television has a 40-inch diagonal with height 24 inches. What is its width?

9. 

10. The diagonal of a square has length $12\sqrt{2}$ ft. Find the length of a side of the square.

10. _______________

288

Chapter 10 RADICAL EXPRESSIONS, EQUATIONS, AND FUNCTIONS

10.8 The Complex Numbers

Learning Objective

a Express imaginary numbers as bi, where b is a nonzero real number, and complex numbers as $a + bi$, where a and b are real numbers.

b Add and subtract complex numbers.

c Multiply complex numbers.

d Write expressions involving powers of i in the form $a + bi$.

e Find conjugates of complex numbers and divide complex numbers.

f Determine whether a given complex number is a solution of an equation.

Key Terms

Use the vocabulary terms listed below to complete each statement in Exercises 1–4.

complex	conjugate	i	imaginary

1. $\sqrt{-1} = $ ___________________ .

2. A(n) ___________________ number is any number that can be written in the form

$a + bi$, where a and b are real numbers and $b \neq 0$.

3. A(n) ___________________ number is any number that can be written in the form

$a + bi$, where a and b are real numbers.

4. The ___________________ of a complex number $a + bi$ is $a - bi$.

Objective a   Express imaginary numbers as bi, where b is a nonzero real number, and complex numbers as $a + bi$, where a and b are real numbers.

Express in terms of i.

5. $\sqrt{-49}$ 5.___________________

6. $\sqrt{-17}$

6.________________

7. $\sqrt{-200}$

7.________________

8. $-\sqrt{-45}$

8.________________

9. $8-\sqrt{-150}$

9.________________

10. $\sqrt{-32}-\sqrt{-9}$

10.________________

Objective b   Add and subtract complex numbers.

Add or subtract and simplify.

11. $(3+5i)+(9-2i)$

11.________________

12. $(5-7i)+(11+4i)$

12.________________

13. $(3-8i)-(6-2i)$

13.________________

290

14. $(-2-5i)-(9-7i)$ 14._______________

Objective c   Multiply complex numbers.

Multiply.

15. $8i \cdot 11i$ 15._______________

16. $\sqrt{-100} \cdot \sqrt{-81}$ 16._______________

17. $2i(5-4i)$ 17._______________

18. $-3i(6-i)$ 18._______________

19. $(-2+3i)(5+6i)$ 19._______________

20. $(7-3i)(3-4i)$

20._____________

21. $(4-7i)^2$

21._____________

22. $(10+3i)^2$

22._____________

Objective d   Write expressions involving powers of i in the form $a + bi$.

Simplify.

23. i^{44}

23._____________

24. $(-i)^{10}$

24._____________

25. $(2i)^3$

25._____________

26. $(2i)^5$

26._____________

292

Objective e   Find conjugates of complex numbers and divide complex numbers.

Divide and simplify to the form a + bi.

27. $\dfrac{3}{2+i}$ 27._______________

28. $\dfrac{11}{5i}$ 28._______________

29. $\dfrac{6-5i}{2i}$ 29._______________

30. $\dfrac{3+2i}{5-3i}$ 30._______________

Determine whether the complex number is a solution of the equation..

31. $1 - 3i$;

    $x^2 - 2x + 10 = 0$

31. _______________

32. $1 + 3i$;

    $x^2 - 2x + 10 = 0$

32. _______________

33. $3 + 4i$;

    $x^2 - 6x + 13 = 0$

33. _______________

Chapter 11 QUADRATIC EQUATIONS AND FUNCTIONS

11.1 The Basics of Solving Quadratic Equations

Learning Objectives

a   Solve quadratic equations using the principle of square roots and find the x-intercepts of the graph of a related function.

b   Solve quadratic equations by completing the square.

c   Solve applied problems using quadratic equations.

Key Terms

Use the vocabulary terms listed below to complete each statement in Exercises 1–4.

    complete   **quadratic**   **standard**

1.   The equation $x^2 = 6 + x$ is an example of a(n) _____________________ equation.

2.   The _____________________ form of a quadratic equation is $ax^2 + bx + c = 0$.

3.   To _____________________ the square for $x^2 + bx$, we add $\left(\dfrac{b}{2}\right)^2$.

Objective a   Solve quadratic equations using the principle of square roots and find the x-intercepts of the graph of a related function.

4.   **a)** Solve: $4x^2 = 24$.   4. a)_______________

    b) Find the x-intercepts of $f(x) = x^2 - 24$.

                                                 b)_______________

5.   **a)** Solve: $4x^2 = -81$.   5. a)_______________

    b) Find the x-intercepts of $f(x) = 4x^2 - 81$.

                                                     b)_______________

Solve. Give the exact solution and approximate solutions to three decimal places, when appropriate.

6. $6t^2 = 18$

6. _________________

7. $4x^2 + 81 = 0$

7. _________________

8. $(y-3)^2 = 100$

8. _________________

9. $(x-5)^2 = 2$

9. _________________

10. $(a+7)^2 = -4$

10. _________________

11. $\left(y + \dfrac{2}{3}\right)^2 = \dfrac{8}{9}$

11. _________________

Objective b Solve quadratic equations by completing the square.

Solve by completing the square. Show your work.

12. $x^2 + x = 20$

12. _________________

13. $t^2 - 6t = -8$

13. _________________

296

14. $t^2 - 4t + 2 = 0$ **14.** _________________

15. $x^2 + 6x - 2 = 0$ **15.** _________________

Complete the square to find the x-intercepts of each function.

16. $f(x) = x^2 + 8x + 9$ **16.** _________________

17. $g(x) = x^2 - 4x + 2$ **17.** _________________

18. $f(x) = 2x^2 + x - 5$ **18.** _________________

19. $f(x) = 4x^2 - 2x - 1$ **19.** _________________

Solve by completing the square. Show your work.

20. $8x^2 + 3 = 10x$ **20.** _________________

21. $3x^2 + 2x - 4 = 0$

21. ___________

22. $x^2 + x + 8 = 0$

22. ___________

23. $x^2 + 8x + 25 = 0$

23. ___________

Objective c   Solve applied problems using quadratic equations.

24. LeBron James has a vertical leap of 44 in. (*Source*: HoopsVibe.com). Use $V = 48T^2$ to estimate his hang time.

24. ___________

The function $s(t) = 16t^2$ is used to approximate the distance s, in feet, that an object falls freely from rest in t seconds. Use the formula for Exercises 25 and 26.

25. The Nurek Dam in Tajikistan, at 984 ft, is the world's tallest dam. How long would it take an object to fall freely from the top?

25. ___________

26. The Spire of Dublin in Ireland, at 393 ft, is the world's tallest sculpture. How long would it take an object to fall freely from the top?

26. ___________

298

Chapter 11 QUADRATIC EQUATIONS AND FUNCTIONS

11.2 The Quadratic Formula

> **Learning Objective**
> a   Solve quadratic equations using the quadratic formula, and approximate solutions using a calculator.

Key Terms

In Exercises 1 and 2, complete each statement.

1. The principle of square roots states that for any real number k, if $x^2 = k$, then

   $x =$ _________________.

2. The quadratic formula states that the solutions of $ax^2 + bx + c = 0, a \neq 0$, are given by

   $x =$ _________________.

Objective a   **Solve quadratic equations using the quadratic formula, and approximate solutions using a calculator.**

Solve.

3. $x^2 + 3x - 2 = 0$ 3. _______________

4. $4a^2 = 32a - 20$ 4. _______________

5. $x^2 + x + 3 = 0$ 5. _______________

6. $t^2 + 11 = 4t$

6. _______________

7. $\dfrac{1}{x^2} - 2 = \dfrac{8}{x}$

7. _______________

8. $4s + s(s - 3) = 1$

8. _______________

9. $10x^2 + 5x = 2$

9. _______________

300

10. $100x^2 - 60x + 9 = 0$ **10.** ______________

11. $6x(x+1) + 3 = 5x(x+2)$ **11.** ______________

12. $5t^2 = 12t + 16$ **12.** ______________

13. $x^2 + 25 = 6x$ **13.** ______________

14. $x^3 - 125 = 0$ **14.** ______________

Solve. Give the exact solutions and approximate solutions to three decimal places.

15. $x^2 + 2x - 5 = 0$

15. _______________

16. $x^2 - 5x + 2 = 0$

16. _______________

17. $2x^2 + 4x - 7 = 0$

17. _______________

18. $3x^2 - 5x - 4 = 0$

18. _______________

302

Chapter 11 QUADRATIC EQUATIONS AND FUNCTIONS

11.3 Applications Involving Quadratic Equations

Learning Objectives
a Solve applied problems involving quadratic equations.
b Solve a formula for a given letter.

Objective a   Solve applied problems involving quadratic equations.

Solve.

1. The width of a rectangular flower garden is 5 feet
    less than the length. The area is 36 ft^2. Find the
    length and width.

    1. _________________

2. The length of a rectangular parking lot is three
    times the width. The area is 1728 ft^2. Find the
    length and the width.

    2. _________________

3. The base of a triangular sail is 4 ft less than its
    height. The area is 72 ft^2. Find the base and the
    height of the sail.

    3. _________________

4. The hypotenuse of a right triangle is 75 m long. The length of one leg is 51 m less than the other. Find the length of the legs.

4. ______________________

5. A student opens a geology book to two facing pages. The product of the page numbers is 5852. Find the page numbers.

5. ______________________

6. The outside of a mosaic mirror frame measures 21 in. by 30 in., and 360 in^2 of mirror shows. Find the width of the frame.

6. ______________________

7. During the first part of a trip, Kai drove 150 mi at a certain speed. Kai then drove another 90 mi at a speed that was 5 mph slower. If the total time of Kai's trip was 5 hr, what was her speed on each part of the trip?

7. ______________________

304

8. Daquan traveled 300 mi averaging a certain speed.
If the car had gone 10 mph slower, the trip would
have taken 1 hr longer. Find Daquan's average speed.

8. _________________

9. Tim flies 500 mi at a certain speed. David flies
1000 mi at a speed that is 75 mph faster, but takes
1 hr longer. Find the speed of each plane.

9. _________________

10. Hwae bikes the 48 mi to the state park averaging
a certain speed. The return trip is made at a speed
that is 2 mph slower. Total time for the round trip
is 14 hr. Find Hwae's average speed on each part
of the trip.

10. _________________

11. The White River flows at a rate of 4 mph. A boat
travels 20 mi upriver and returns in a total time of
3 hr. What is the speed of the boat in still water?

11. _________________

Objective b   Solve a formula for a given letter.

Solve each formula for the given letter. Assume that all variables represent nonnegative numbers.

12. $T = 8v^2$, for v

12. _______________

13. $y = 3x^2 + 5x$, for x

13. _______________

14. $y = \dfrac{kxz}{w^2}$, for w

14. _______________

15. $M = 5\sqrt{\dfrac{p}{q}}$, for q

15. _______________

16. $a^2 = b^2 + c^2 - 2abX$, for c

16. _______________

17. $t = x_0 v + \dfrac{av^2}{8}$, for v

17. _______________

18. $B = \dfrac{1}{3}(x^2 - 2x)$, for x

18. _______________

19. $H = 1.6\sqrt{d}$, for d

19. _______________

Chapter 11 QUADRATIC EQUATIONS AND FUNCTIONS

11.4 More on Quadratic Equations

Learning Objectives
a   Determine the nature of the solutions of a quadratic equation.
b   Write a quadratic equation having two given numbers as solutions.
c   Solve equations that are quadratic in form.

Key Terms
Use the vocabulary terms listed below to complete each statement in Exercises 1–4.

conjugates   **discriminant**   **real**

1.   The expression $b^2 - 4ac$ in the quadratic formula is called the _______________________.

2.   When $b^2 - 4ac = 0$ there is one _______________ number solution.

3.   When $b^2 - 4ac$ is negative, the solutions are complex _______________.

Objective a   Determine the nature of the solutions of a quadratic equation.

Determine the nature of the solutions of each equation.

4. $x^2 - 3x + 10 = 0$ 4. _______________

5. $x^2 - 12 = 0$ 5. _______________

6. $t^2 + 15 = 0$ 6. _______________

7. $6x^2 - 5x = 0$ 7. _______________

8. $6x^2 - 19x + 10 = 0$ 8. _______________

9. $a^2 + a + 4 = 0$

9. ________________

10. $16x^2 + 24x + 9 = 0$

10. ________________

11. $3x^2 - 7x = -2$

11. ________________

Objective b Write a quadratic equation having two given numbers as solutions.

Write a quadratic equation having the given numbers as solutions.

12. $-5, 6$

12. ________________

13. 4, only solution

13. ________________

14. $-2, -10$

14. ________________

15. $4, -\dfrac{3}{5}$

15. ________________

16. $-\sqrt{5}, \sqrt{5}$

16. ________________

17. $3\sqrt{10}, -3\sqrt{10}$

17. ________________

18. $5i, -5i$

18. ________________

308

Objective c   Solve equations that are quadratic in form.

Solve.

19. $x^4 - 17x^2 + 16 = 0$

19. _________________

20. $a^4 - 17a^2 + 72 = 0$

20. _________________

21. $t - 3\sqrt{t} - 10 = 0$

21. _________________

22. $\left(m^2 - 5\right)^2 - 8\left(m^2 - 5\right) + 15 = 0$

22. _________________

23. $m^{-2} + m^{-1} - 30 = 0$

23. _______________

24. $6x^{-2} - 19x^{-1} + 10 = 0$

24. _______________

25. $w^{2/5} - 3w^{1/5} - 4 = 0$

25. _______________

Find the x-intercepts of the graph of each function.

26. $f(x) = 3x + 13\sqrt{x} - 10$

26. _______________

27. $f(x) = (x^2 - 2x)^2 - 11(x^2 - 2x) + 24$

27. _______________

28. $f(x) = x^{2/3} - x^{1/3} - 20$

28. _______________

310

Chapter 11 QUADRATIC EQUATIONS AND FUNCTIONS

11.5 Graphing $f(x) = a(x - h)^2 + k$

Learning Objective

a   Graph quadratic functions of the type $f(x) = ax^2$ and then label the vertex and the line of symmetry.

b   Graph quadratic functions of the type $f(x) = a(x - h)^2$ and then label the vertex and the line of symmetry.

c   Graph quadratic functions of the type $f(x) = a(x - h)^2 + k$, finding the vertex, the line of symmetry, and the maximum or minimum function value, or y-value.

Key Terms

Use the vocabulary terms listed below to complete each statement in Exercises 1–4.

     axis of symmetry   **parabola**   **translated**   **vertex**

1.   The graph of a quadratic equation is a(n) ____________________.

2.   The graph of a quadratic equation is symmetric with respect to its

     ____________________.

3.   The maximum or minimum value of a quadratic function occurs at the

     ____________________ of its graph.

4.   The graph of $f(x) = a(x - h)^2 + k$ looks like the graph of $f(x) = ax^2$ except that it is

     ____________________ $|h|$ units horizontally and $|k|$ units vertically.

Objective a   **Graph quadratic functions of the type $f(x) = ax^2$ and then label the vertex and the line of symmetry.**

Objective b   **Graph quadratic functions of the type $f(x) = a(x - h)^2$ and then label the vertex and the line of symmetry.**

For each of the following, graph the function, label the vertex, and draw the axis of symmetry.

5. $f(x) = (x+2)^2$

5.

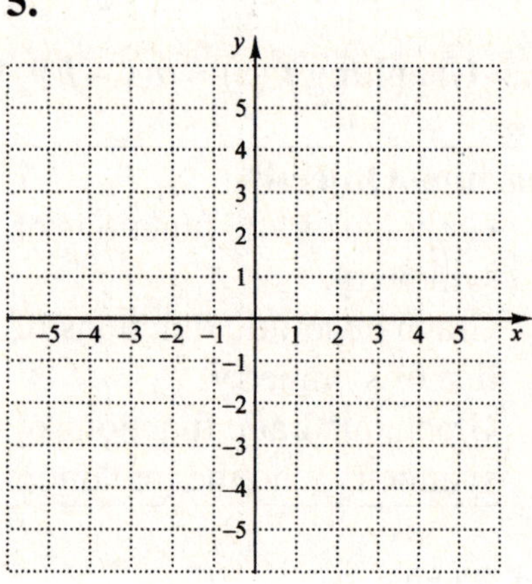

6. $f(x) = (x-6)^2$

6.

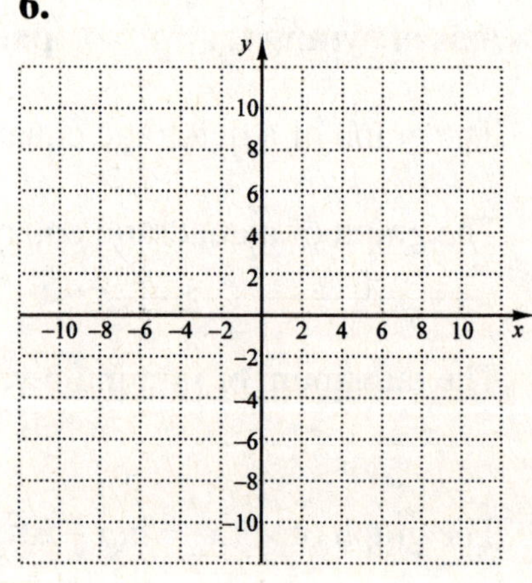

7. $g(x) = -(x-3)^2$

7.

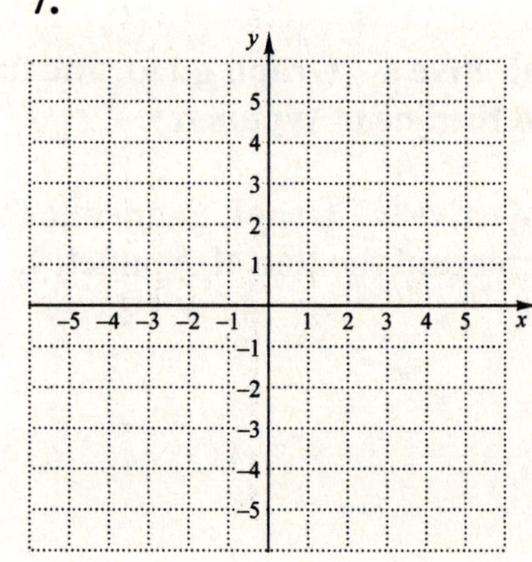

312

8. $g(x) = -2(x+1)^2$

8.

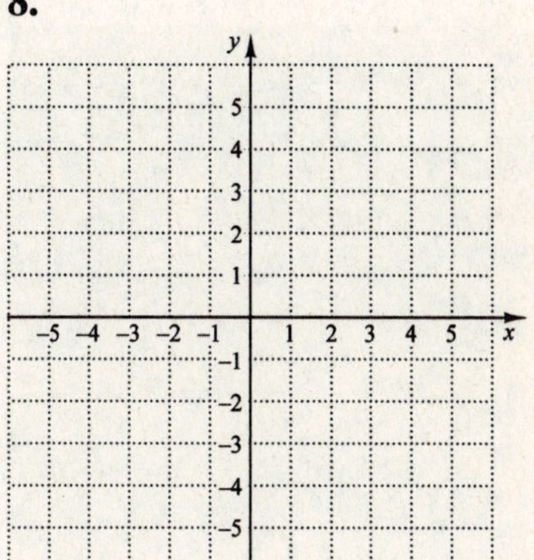

Objective c **Graph quadratic functions of the type** $f(x) = a(x-h)^2 + k$**, finding the vertex, the line of symmetry, and the maximum or minimum function value, or** y**-value.**

For each of the following, graph the function and find the vertex, the axis of symmetry, and the maximum value or the minimum value.

9. $f(x) = (x+2)^2 - 3$

9. 

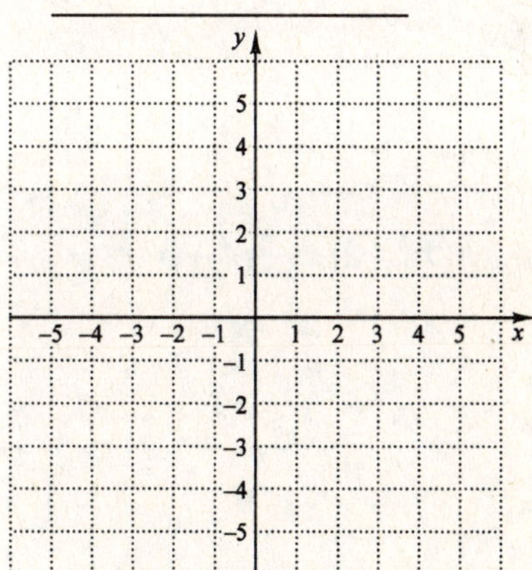

313

10. $f(x) = -(x-2)^2 + 1$

10.___________________

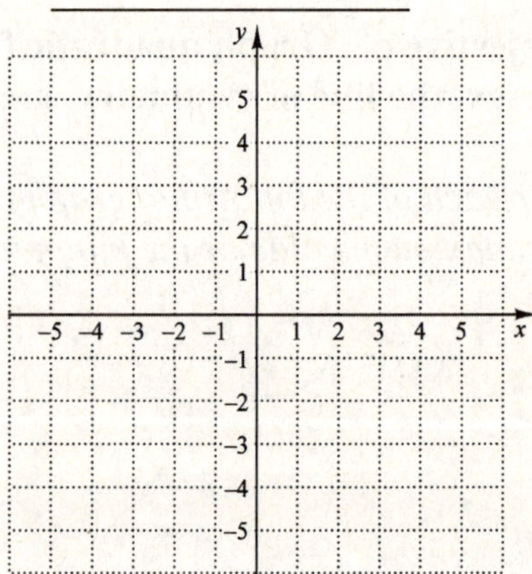

11. $h(x) = -\dfrac{1}{2}(x-3)^2 + 4$

11.___________________

12. $h(x) = 2(x-4)^2 - 3$

12.___________________

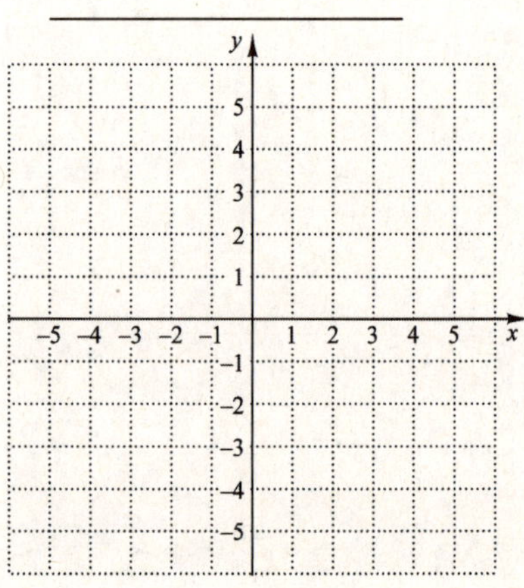

314

Chapter 11 QUADRATIC EQUATIONS AND FUNCTIONS

11.6 Graphing $f(x) = ax^2 + bx + c$

Learning Objective
a For a quadratic function, find the vertex, the line of symmetry, and the maximum or minimum value, and then graph the function.
b Find the intercepts of a quadratic function.

Key Terms
Use the vocabulary terms listed below to complete each statement in Exercises 1–4.

maximum	**minimum**	**x-intercept**	**y-intercept**

1. The function given by $f(x) = a(x-h)^2 + k$ has a(n) ________________ value of k if a is positive.

2. The function given by $f(x) = a(x-h)^2 + k$ has a(n) ________________ value of k if a is negative.

3. For any function, the ________________ occurs at $f(0)$.

4. A(n) ________________ will occur at an input for which $f(x) = 0$.

Objective a   For a quadratic function, find the vertex, the line of symmetry, and the maximum or minimum value, and then graph the function.

For each quadratic function, **a)** *write the function in the form* $f(x) = a(x-h)^2 + k$ *and* **b)** *find the vertex and the axis of symmetry.*

5. $f(x) = x^2 + 8x - 5$ 5. a)________________

                                                    b)________________

*For each quadratic function, find **a)** the vertex and the axis of symmetry, **b)** the minimum or maximum value, and **c)** graph the function.*

6. $f(x) = x^2 + 2x + 3$

6. a)_________________

b)_________________

c)

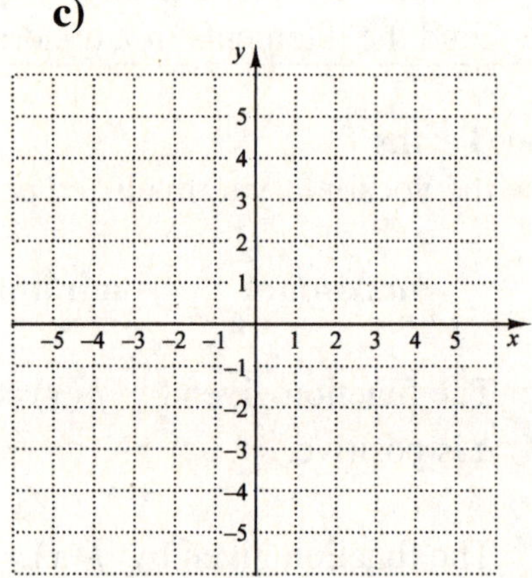

7. $f(x) = x^2 - 4x - 1$

7. a)_________________

b)_________________

c)

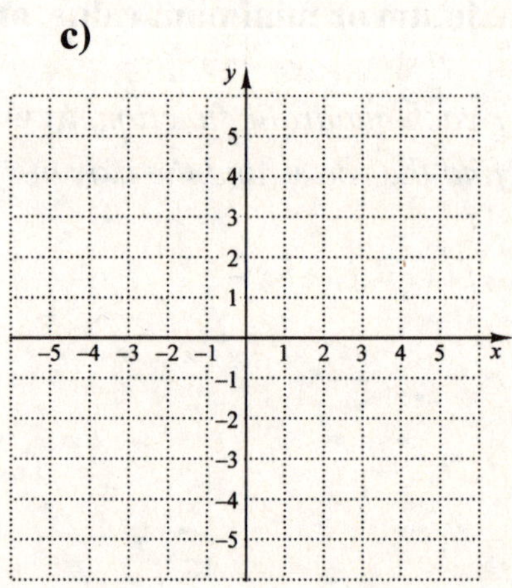

316

8. $h(x) = 2x^2 - 8x + 9$

8. a) ______________________

   b) ______________________

   c)

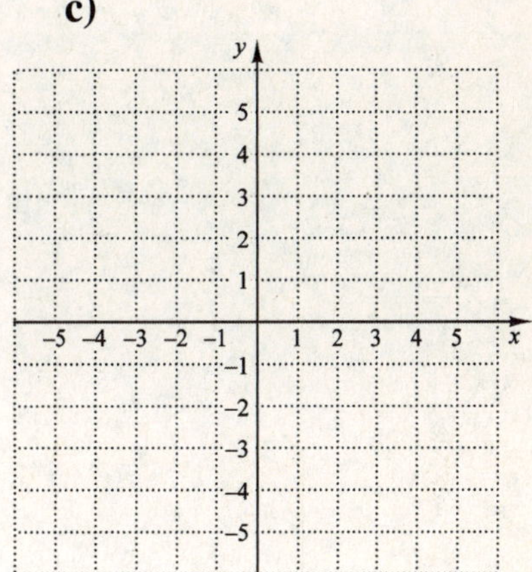

9. $f(x) = -x^2 + 2x - 5$

9. a) ______________________

   b) ______________________

   c)

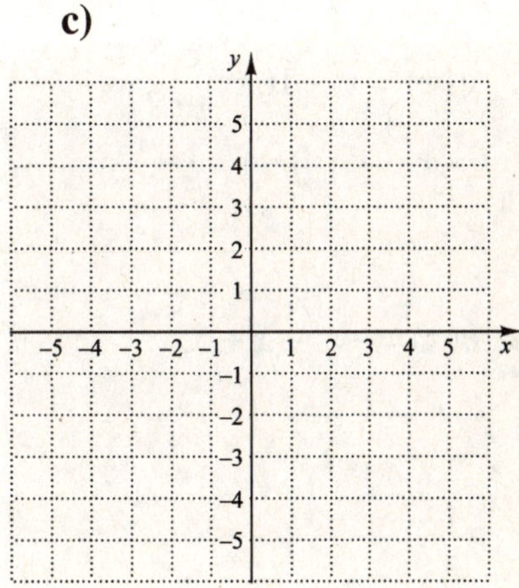

10. $f(x) = -\frac{1}{2}x^2 + 2x + 1$

10. a)___________________

b)___________________

c)

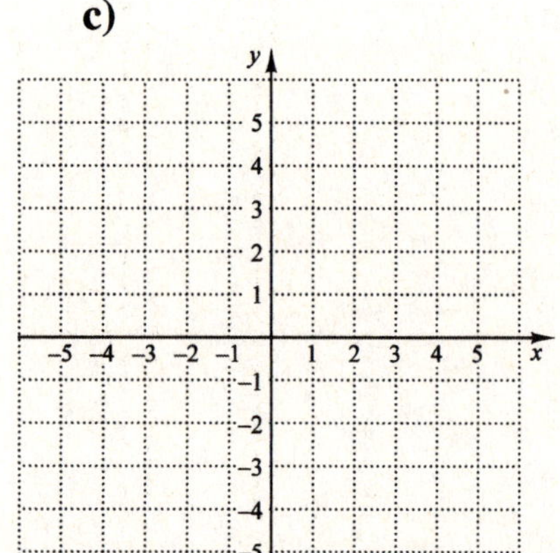

Objective b   Find the intercepts of a quadratic function.

Find the x- and y-intercepts.

11. $f(x) = x^2 + 6x - 1$

11. ___________________

12. $g(x) = x^2 - 10x$

12. ___________________

13. $h(x) = -x^2 + x + 6$

13. ___________________

14. $f(x) = 2x^2 - 3x + 8$

14. ___________________

318

Chapter 11 QUADRATIC EQUATIONS AND FUNCTIONS

11.7 Mathematical Modeling with Quadratic Functions

> **Learning Objectives**
> a   Solve maximum-minimum problems involving quadratic functions.
> b   Fit a quadratic function to a set of data to form a mathematical model, and solve related
>   applied problems.

Objective a   Solve maximum-minimum problems involving quadratic functions.

Solve.

1. The value of a share of stock can be represented   1. _______________
    by $f(x) = x^2 - 4x + 10$, where x is the number of
    months after January 2007. What is the lowest value
    $f(x)$ will reach, and when will that occur?

2. The average cost per box built at Green Box   2. _______________
    Company can be estimated by
    $c(x) = 0.1x^2 - 0.6x + 1.35$, where $c(x)$ is in dollars.
    What is minimum average cost per box and how
    many boxes should be built to achieve that minimum?

3. A rectangular garden has a perimeter of 84 ft.   3. _______________
    What dimensions yield the maximum area?

4. Nick is fencing in a rectangular section of his yard. He has 90 ft of fence, and will use a wall of his house as one side of the rectangle. What is the maximum area Nick can enclose? What should the dimensions of the rectangle be in order to yield this area?

4. _______________

5. A raised rectangular garden is formed in a corner of a fenced yard, with 10 ft of treated lumber completing the other two sides of the rectangle. If the lumber is 8 in. high, what dimensions of the base will maximize the garden's volume?

5. _______________

6. What is the maximum product of two numbers that add to 24? What are the numbers?

6. _______________

7. What is the minimum product of two numbers that differ by 10? What are the numbers?

7. _______________

320

8. What is the maximum product of two numbers that
add to -14? What numbers yield this product?

8. _______________

**Objective b   Fit a quadratic function to a set of data to form a mathematical model,
and solve related applied problems.**

*For the scatterplots and graphs in Exercises 9–12, determine which, if any, of the following
functions might be used as a model for the data: Linear, $f(x) = mx + b$; quadratic,*

$f(x) = ax^2 + bx + c, a > 0$; *quadratic,* $f(x) = ax^2 + bx + c, a < 0$; *polynomial, neither
quadratic nor linear.*

9.

9. _______________

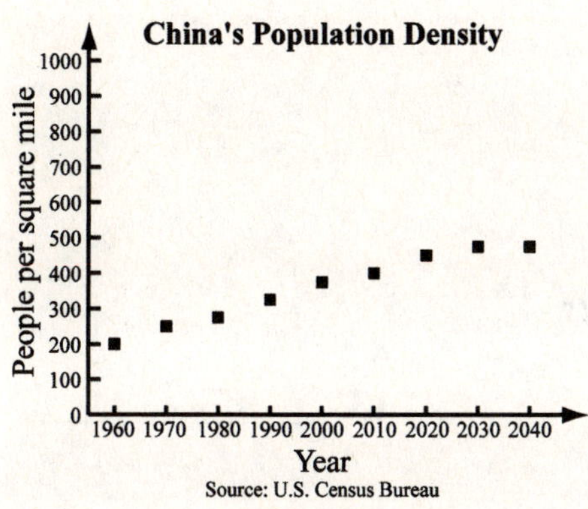

10.

10. _______________

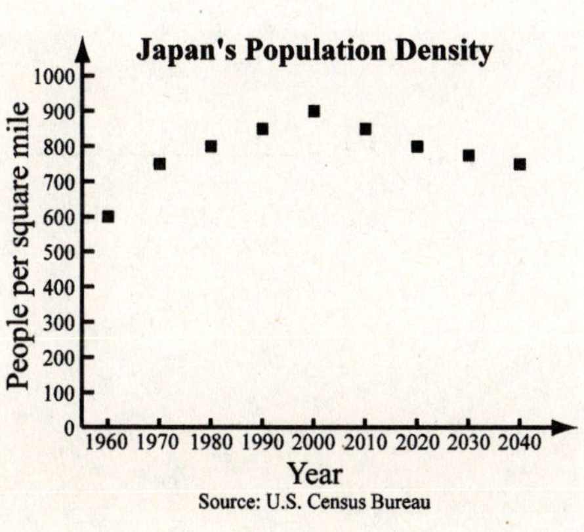

321

11.

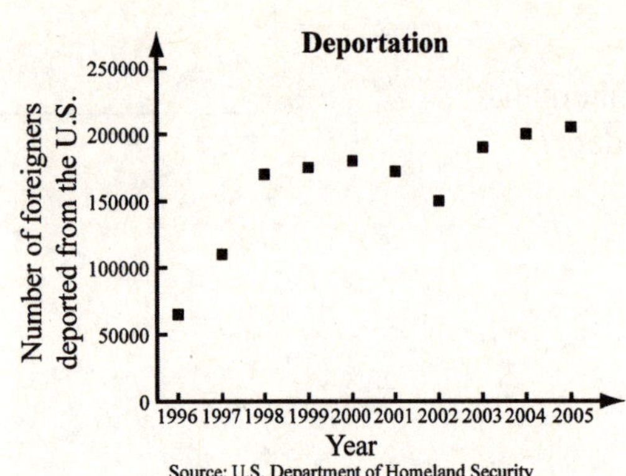

11. _______________

12.

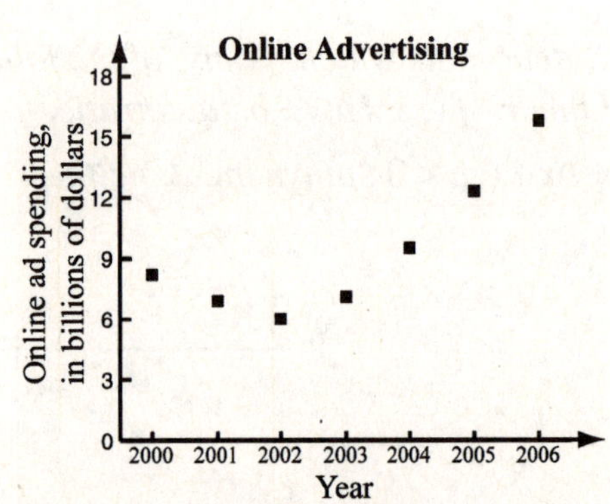

12. _______________

Find a quadratic function that fits the set of data points.

13. $(1, 3), (-1, -1), (2, 10)$

13. _______________

14. $(1, 0), (-1, 8), (2, 6)$

14. _______________

322

Chapter 11 QUADRATIC EQUATIONS AND FUNCTIONS

11.8 Polynomial Inequalities and Rational Inequalities

Learning Objective
a   Solve quadratic inequalities and other polynomial inequalities.
b   Solve rational inequalities.

Key Terms
Use the vocabulary terms listed below to complete each statement in Exercises 1–4.

       polynomial   **rational**   **test points**   **zeros**

1.   An inequality like $x^3 - x > 6x^2 + 7$ is an example of a ___________________ inequality.

2.   The ___________________ of a function occur at the x-intercepts of the graph of the function.

3.   We use ___________________ to determine the sign of a polynomial over an interval of the x-axis.

4.   An inequality like $\dfrac{x+4}{2x+1} \le 0$ is an example of a ___________________ inequality.

Objective a   Solve quadratic inequalities and other polynomial inequalities.

Solve algebraically and verify results from the graph.

5. $(x+1)(2x-9) < 0$ 5. _______________

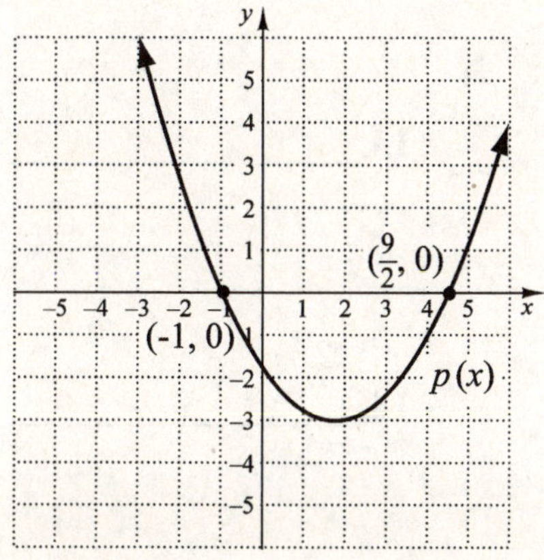

Solve.

6. $(x+2)(x-5) > 0$

6. ______________

7. $(x-4)(x+1) \geq 0$

7. ______________

8. $x^2 + x - 6 \leq 0$

8. ______________

9. $x^2 + 2x + 1 < 0$

9. ______________

10. $5x(x-3)(x+6) < 0$

10. ______________

Objective b   Solve rational inequalities.

Solve.

11. $\dfrac{1}{x-3} > 0$

11. ______________

12. $\dfrac{x+2}{x-1} \leq 4$

12. ______________

13. $\dfrac{(x+3)(x-5)}{x+2} \geq 0$

13. ______________

14. $\dfrac{x}{7-x} > 0$

14. ______________

324

Chapter 12 EXPONENTIAL AND LOGARITHMIC FUNCTIONS

12.1 Exponential Functions

Learning Objectives
a   Graph exponential equations and functions.
b   Graph exponential equations in which x and y have been interchanged.
c   Solve applied problems involving applications of exponential functions and their graphs.

Key Terms
Use the vocabulary terms listed below to complete each statement in Exercises 1–4.

asymptote	**decreasing**	**exponential**	**increasing**

1.   A function in the form $f(x) = a^x$ is a(n) _________________ function.

2.   A function f is _________________ if the values of $f(x)$ increase as x increases.

3.   A function f is _________________ if the values of $f(x)$ decrease as x increases.

4.   The graph of a function $f(x) = a^x$ has the x-axis as a(n) _________________.

Objective a   Graph exponential equations and functions.

Graph.

5. $f(x) = 4^x$

5.

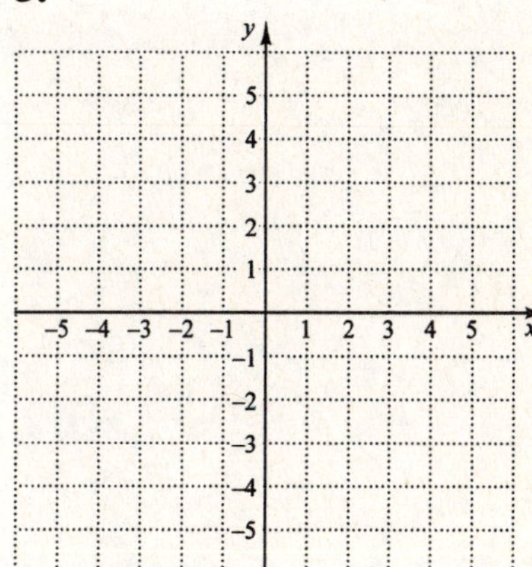

6. $f(x) = 2^{x-3}$

6.

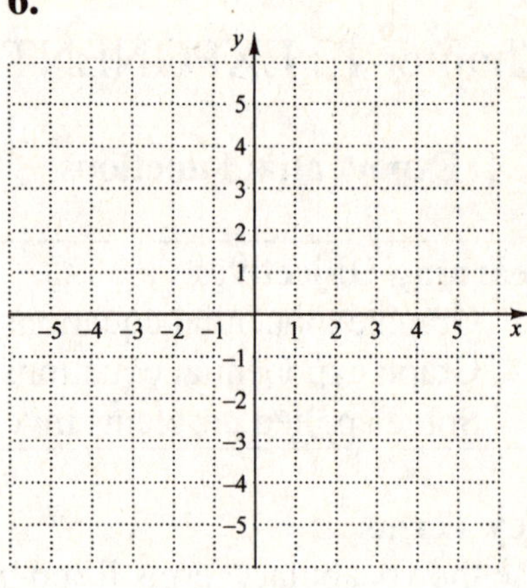

7. $f(x) = \left(\dfrac{1}{8}\right)^x$

7.

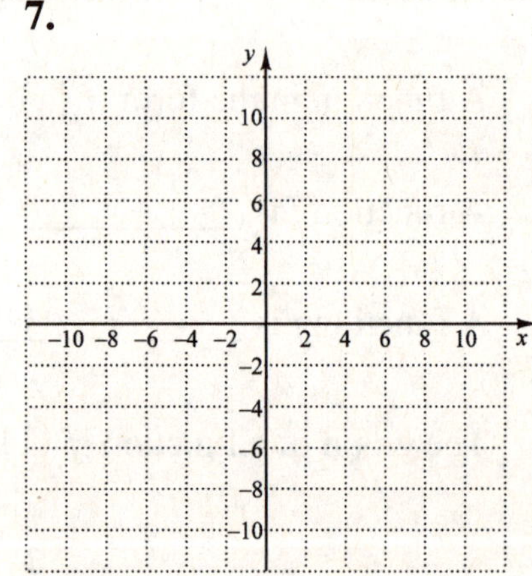

Objective b   Graph exponential equations in which x and y have been interchanged.

Graph.

8. $x = 2^y$

8.

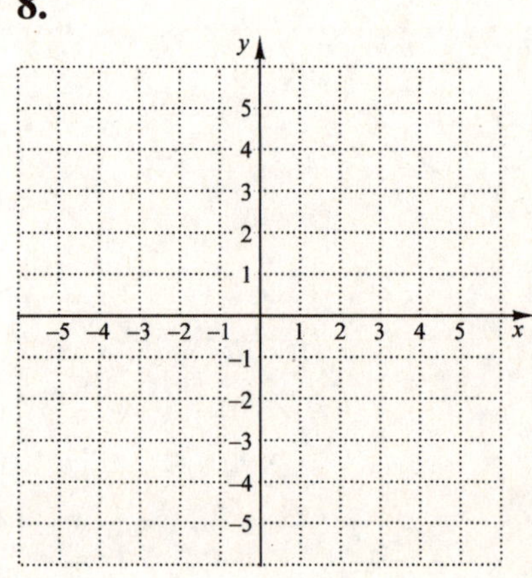

326

9. $x = \left(\dfrac{1}{2}\right)^{y}$

9.

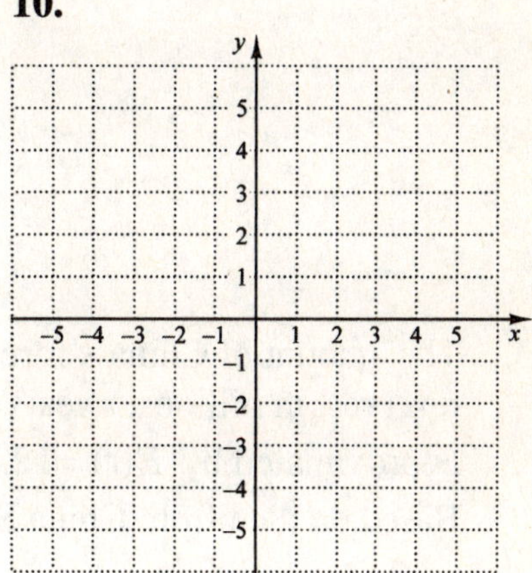

Graph both equations using the same set of axes.

10. $y = 4^{x}$, $x = 4^{y}$

10.

11. $y = \left(\dfrac{1}{3}\right)^{x}$, $x = \left(\dfrac{1}{3}\right)^{y}$

11.

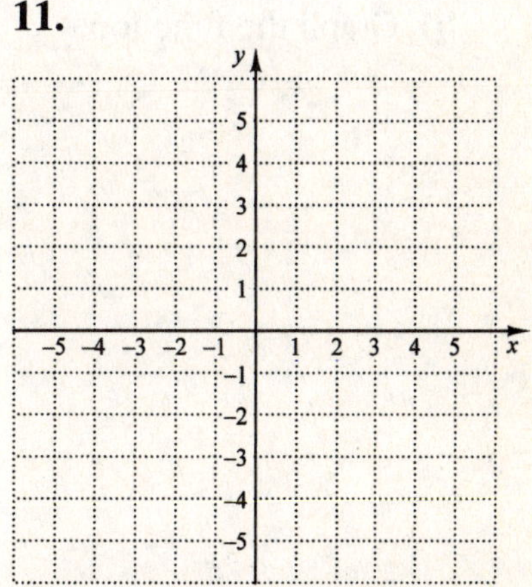

327

Objective c Solve applied problems involving applications of exponential functions and their graphs.

Solve.

12. The population of Ethiopia, in millions, t years after 2003, can be approximated by $P(t) = 66.56(1.221)^t$ (*Source*: Based on data from *Time Almanac 2004*).

   a) Predict the population of Ethiopia in 2008 and in 2010.
   b) Graph the function.

12. a)________________

b)

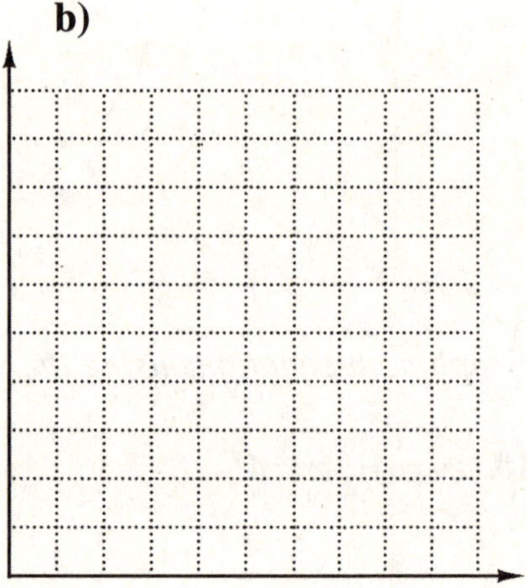

13. The amount of China's foreign-exchange reserves, in billions, t years after 1990, can be approximated by $F(t) = 12.4(1.32)^t$ (*Source*: Based on data from China Statistic Yearbook).

   a) Find the amount of China's foreign-exchange reserves in 1990 and in 2005.
   b) Graph the function.

13. a)________________

b)

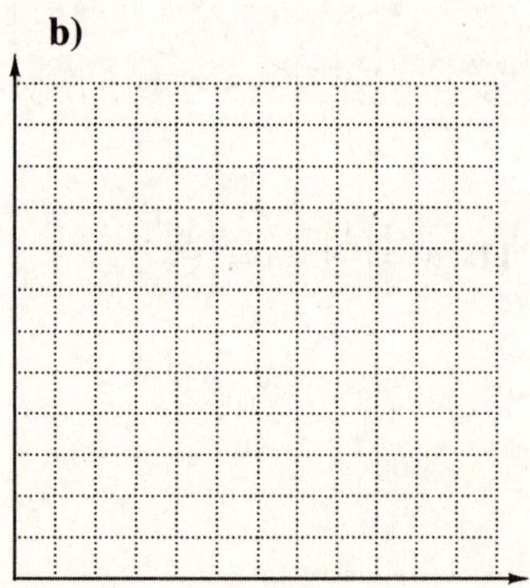

Chapter 12 EXPONENTIAL AND LOGARITHMIC FUNCTIONS

12.2 Inverse Functions and Composite Functions

Learning Objective

a   Find the inverse of a relation if it is described as a set of ordered pairs or as an equation.

b   Given a function, determine whether it is one-to-one and has an inverse that is a function.

c   Find a formula for the inverse of a function, if it exists, and graph inverse relations and functions.

d   Find the composition of functions and express certain functions as a composition of functions.

e   Determine whether a function is an inverse by checking its composition with the original function.

Key Terms

Use the vocabulary terms listed below to complete each statement in Exercises 1–4.

composite	horizontal-line	inverse	one-to-one

1.   In a(n) _________________ function, the output depends on a variable that, in turn, depends on another variable.

2.   A function f is _________________ if different inputs have different outputs.

3.   Exchanging the first and second coordinates of all ordered pairs in a relation results in the _________________ relation.

4.   A function f is one-to-one if it passes the _________________ test.

Objective a   Find the inverse of a relation if it is described as a set of ordered pairs or as an equation.

Find the inverse of the relation.

5. $\{(2, -1), (6, 3), (6, -4), (8, 0)\}$ **5.** _________________

Find an equation of the inverse of the relation. Then complete the second table.

6. $y = 4x + 2$

x	y
−1	−2
0	2
1	6

x	y
−2	
2	
6	

6. ______________________

Objective b   Given a function, determine whether it is one-to-one and has an inverse that is a function.

Determine whether each function is one-to-one.

7. $f(x) = 3x - 4$

7. ______________________

8. $f(x) = x^2 + 5$

8. ______________________

9. $f(x) = |x + 3|$

9. ______________________

10. $f(x) = 5^x$

10. ______________________

Objective c   Find a formula for the inverse of a function, if it exists, and graph inverse relations and functions.

Determine whether each function is one-to-one. If it is, find a formula for its inverse.

11. $f(x) = x + 1$

11. ______________________

12. $f(x) = 5x - 2$

12. ______________________

330

13. $f(x) = x^2 - 3$

13. _______________

14. $f(x) = \dfrac{2}{x}$

14. _______________

15. $f(x) = x^3 + 2$

15. _______________

16. $f(x) = 10$

16. _______________

Graph each function and its inverse on the same set of axes.

17. $f(x) = \dfrac{4}{5}x + 3$

17.

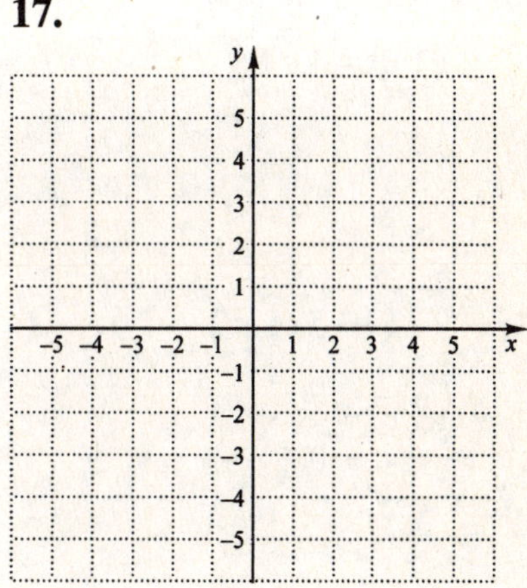

331

18. $f(x) = x^3 - 4$

18.

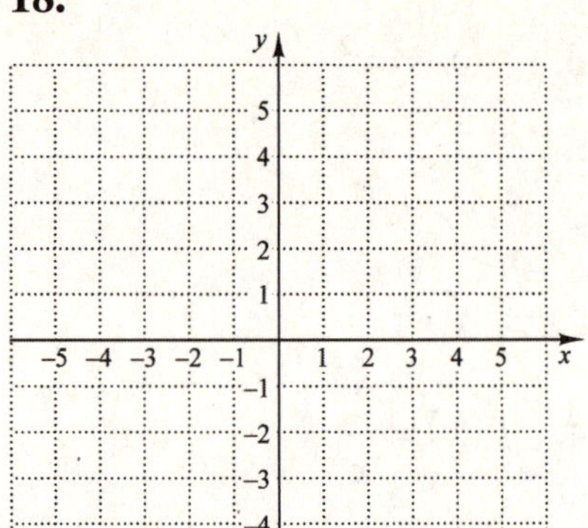

Objective d **Find the composition of functions and express certain functions as a composition of functions.**

Find $(f \circ g)(x)$ *and* $(g \circ f)(x)$.

19. $f(x) = 7 - 3x$

$g(x) = 0.46x + 5$

19._________________

____

20. $f(x) = 2x^2 - 6$

$g(x) = 4x + 9$

20._________________

____

21. $f(x) = 5x^2 + 2$

$g(x) = \dfrac{4}{x}$

21._________________

____

22. $f(x) = x^2 - 4$

$g(x) = x^2 + 4$

22._________________

____

332

Find $f(x)$ and $g(x)$ such that $h(x) = (f \circ g)(x)$. Answers may vary.

23. $h(x) = (10 - 3x)^2$

23._______________

____

24. $h(x) = \sqrt{4x + 9}$

24._______________

____

25. $h(x) = \dfrac{4}{\sqrt{2x + 7}}$

25._______________

____

26. $h(x) = \dfrac{1}{\sqrt{5x}} - \sqrt{5x}$

26._______________

____

Objective e   Determine whether a function is an inverse by checking its composition with the original function.

27. Let $f(x) = \sqrt[3]{x + 2}$. Use composition to show that
$f^{-1}(x) = x^3 - 2$.

27._______________

____

____

____

28. Let $f(x) = \dfrac{1+x}{x}$. Use composition to show that

$$f^{-1}(x) = \dfrac{1}{x-1}.$$

28._________________

____

____

____

Find the inverse of the given function by thinking about the operations of the function and then reversing, or undoing, them. Then use composition to show whether the inverse is correct.

29. $f(x) = 4x$

29._________________

30. $f(x) = \dfrac{1}{5}x - 9$

30._________________

31. $f(x) = \dfrac{1}{x}$

31._________________

32. $f(x) = \sqrt[3]{x+6}$

32._________________

33. Size 39 shoes in France are size 7 in the United States. A function that converts shoe sizes in France to those in the United States is $f(x) = x - 32$.

   a) Find the shoe size in the United States that corresponds to size 37 in France.

   b) Determine whether this function has an inverse that is a function. If so, find a formula for the inverse.

   c) Use the inverse function to find the shoe size in France that corresponds to size 8 in the United States.

33.

a)_______________

b)_______________

c)_______________

334

Chapter 12 EXPONENTIAL AND LOGARITHMIC FUNCTIONS

12.3 Logarithmic Functions

Learning Objectives
a   Graph logarithmic equations.
b   Convert from exponential equations to logarithmic equations and from logarithmic equations to exponential equations.
c   Solve logarithmic equations.
d   Find common logarithms on a calculator.

Key Terms
Use the vocabulary terms listed below to complete each statement in Exercises 1–4.

common   **exponent**   **log** x **logarithmic**

1.   The inverse of an exponential function is a(n) _____________________ function.

2. $\log_2 8$ is the _____________________ to which we raise 2 to get 8.

3.   Base-10 logarithms are called _____________________ logarithms.

4. $\log_{10} x$ is abbreviated _____________________.

Objective a   Graph logarithmic equations.

Graph.

5. $y = \log_5 x$

5.

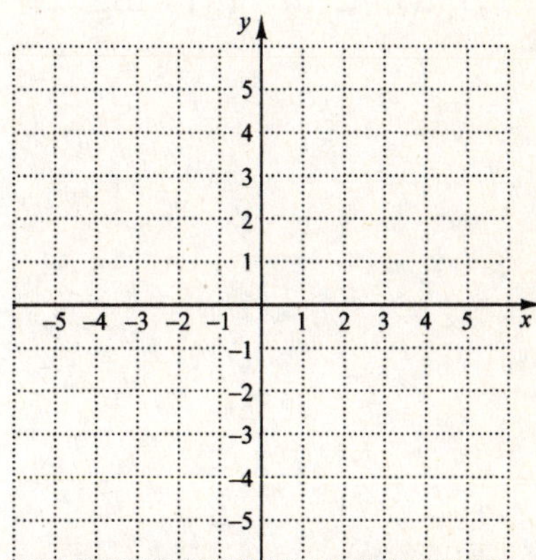

6. $f(x) = \log_{10} x$

6.

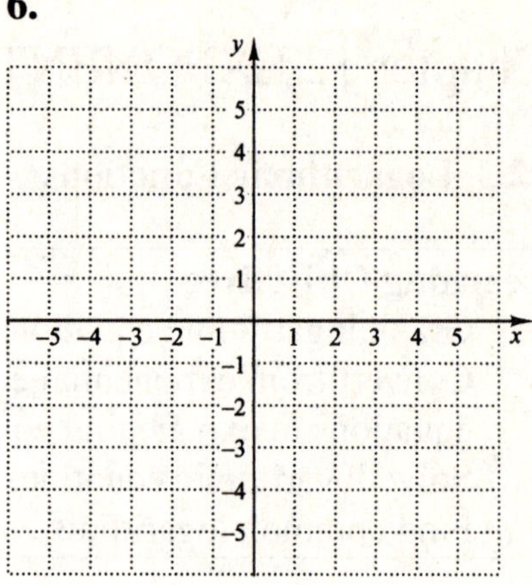

Graph both functions using the same set of axes.

7. $f(x) = 6^x,\ f^{-1}(x) = \log_6 x$

7.

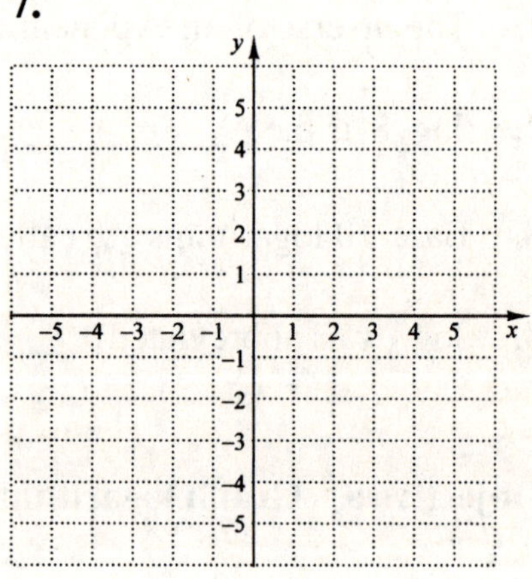

8. $f(x) = 2^x,\ f^{-1}(x) = \log_2 x$

8.

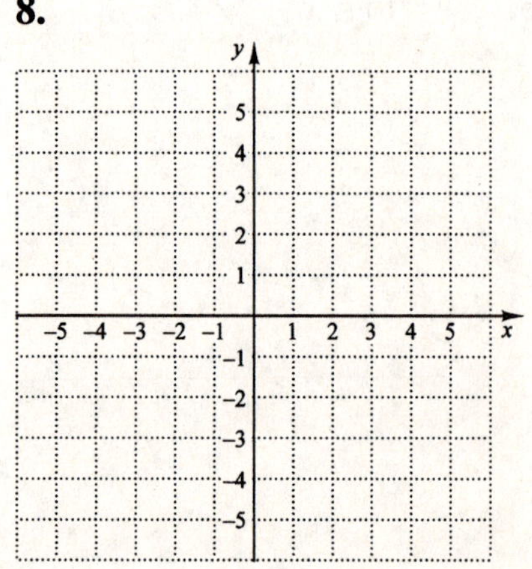

336

Name: ___________ Date: ___________
Instructor: ___________ Section: ___________

Objective b   Convert from exponential equations to logarithmic equations and from logarithmic equations to exponential equations.

Convert to a logarithmic equation.

9. $27^{2/3} = 9$

9. _______________

10. $d^t = m$

10. _______________

11. $e^{-3} = 0.0498$

11. _______________

Convert to an exponential equation.

12. $\log_5 0.2 = -1$

12. _______________

13. $\log_a t = p$

13. _______________

14. $\log_e 0.4 = -0.9163$

14. _______________

Objective c   Solve logarithmic equations.

Solve.

15. $\log_2 x = 4$

15. _______________

16. $\log_5 125 = x$

16. _______________

17. $\log_x 1000 = 1$

17. _______________

337

18. $\log_{81} x = \frac{1}{2}$

18. _________________

Find each of the following.

19. $\log_{10} 100,000$

19. _________________

20. $\log_4 64$

20. _________________

21. $\log_8 \frac{1}{8}$

21. _________________

22. $\log_9 9$

22. _________________

23. $\log_5 1$

23. _________________

Objective d   Find common logarithms on a calculator.

Find the common logarithm, to four decimal places, on a calculator.

24. $\log 87,742.5$

24. _________________

25. $\log 0.29$

25. _________________

26. $\log(-41)$

26. _________________

27. $\log \dfrac{15}{36.2}$

27. _________________

338

Chapter 12 EXPONENTIAL AND LOGARITHMIC FUNCTIONS

12.4 Properties of Logarithmic Functions

Learning Objectives

a   Express the logarithm of a product as a sum of logarithms, and conversely.
b   Express the logarithm of a power as a product.
c   Express the logarithm of a quotient as a difference of logarithms, and conversely.
d   Convert from logarithms of products, quotients, and powers to expressions in terms of individual logarithms, and conversely.
e   Simplify expressions of the type $\log_a a^k$.

Key Terms

In Exercises 1–6, match the expression with an equivalent expression from the column on the right.

1. _______ $\log_a (MN)$ a) k

2. _______ $\log_a M^p$ b) 0

3. _______ $\log_a \dfrac{M}{N}$ c) $\log_a M + \log_a N$

4. _______ $\log_a a^k$ d) $p \cdot \log_a M$

5. _______ $\log_a a$ e) $\log_a M - \log_a N$

6. _______ $\log_a 1$ f) 1

Objective a   Express the logarithm of a product as a sum of logarithms, and conversely.

Express as a sum of logarithms.

7. $\log_2 (4 \cdot 128)$ 7. _______________

8. $\log_b (5xy)$ 8. _______________

Express as a single logarithm.

9. $\log_a 10 + \log_a 9$

9. _______________

10. $\log_y A + \log_y B$

10. _______________

Objective b   **Express the logarithm of a power as a product.**

Express as a product

11. $\log_a t^6$

11. _______________

12. $\log_c H^{-5}$

12. _______________

Objective c   **Express the logarithm of a quotient as a difference of logarithms, and conversely.**

Express as a difference of logarithms.

13. $\log_3 \dfrac{10}{13}$

13. _______________

14. $\log_t \dfrac{u}{v}$

14. _______________

Express as a single logarithm.

15. $\log_a 25 - \log_a 11$

15. _______________

16. $\log_b 48 - \log_b 4$

16. _______________

340

Objective d   Convert from logarithms of products, quotients, and powers to expressions in terms of individual logarithms, and conversely.

Express in terms of logarithms of a single variable or a number.

17. $\log_b(wxyz)$

17. ______________

18. $\log_a(x^{-2}yz^3)$

18. ______________

19. $\log_a \dfrac{w^3 x}{yz^2}$

19. ______________

20. $\log_a \sqrt[4]{\dfrac{x^{20}y^4}{a^5 z^{11}}}$

20. ______________

Express as a single logarithm and, if possible, simplify.

21. $5\log_x p + 4\log_x q$

21. ______________

22. $\log_b bt^2 - 3\log_b \sqrt[3]{t}$

22. ______________

23. $\log_a 3y + 5(\log_a y - \log_a z)$

23. ______________

24. $\log_a (5x-5) - \log_a \left(x^2 - 1\right)$

24. ______________

Given $\log_a 2 = 0.356$ and $\log_a 5 = 0.827$, find each of the following.

25. $\log_a 10$

25. ______________

26. $\log_a \dfrac{5}{2}$

26. ______________

27. $\log_a \dfrac{1}{2}$

27. ______________

28. $\log_a 7$

28. ______________

Objective e   Simplify expressions of the type $\log_a a^k$.

Simplify.

29. $\log_b b^{19}$

29. ______________

30. $\log_e e^p$

30. ______________

Solve for x.

31. $\log_7 7^6 = x$

31. ______________

32. $\log_a a^x = 4.7$

32. ______________

342

Chapter 12 EXPONENTIAL AND LOGARITHMIC FUNCTIONS

12.5 Natural Logarithmic Functions

Learning Objective
a   Find logarithms or powers, base e, using a calculator.
b   Use the change-of-base formula to find logarithms with bases other than e or 10.
c   Graph exponential and logarithmic functions, base e.

Key Terms
Use the vocabulary terms listed below to complete each statement in Exercises 1–4.

change-of-base   **domain**   **natural**   **range**

1. Logarithms base e are called _________________ logarithms.

2. To find a logarithm with a base other than 10 or e, we use the _________________ formula.

3. The _________________ of $f(x) = e^x$ is $\mathbb{R}$.

4. The _________________ of $f(x) = e^x$ is $(0, \infty)$.

Objective a   Find logarithms or powers, base e, using a calculator.

Find each of the following logarithms or powers, base e, using a calculator. Round answers to four decimal places.

5. $\ln 8$ 5. _________________

6. $\ln 0.1$ 6. _________________

7. $\ln\left(\dfrac{2500}{15}\right)$ 7. _________________

343

8. $\ln 0$

8. _______________

9. $\ln e^3$

9. _______________

10. $\ln 35$

10. _______________

11. $\ln 0.0052$

11. _______________

12. $e^{-3.8}$

12. _______________

13. $e^{0.587}$

13. _______________

Objective b   **Use the change-of-base formula to find logarithms with bases other than e or 10.**

Find each of the following logarithms using the change-of-base formula.

14. $\log_4 85$

14. _______________

15. $\log_5 100$

15. _______________

344

16. $\log_{0.2} 90$

16. _______________

17. $\log_{\pi} 32$

17. _______________

Objective c   Graph exponential and logarithmic functions, base e.

Graph.

18. $f(x) = e^x + 1$

18.

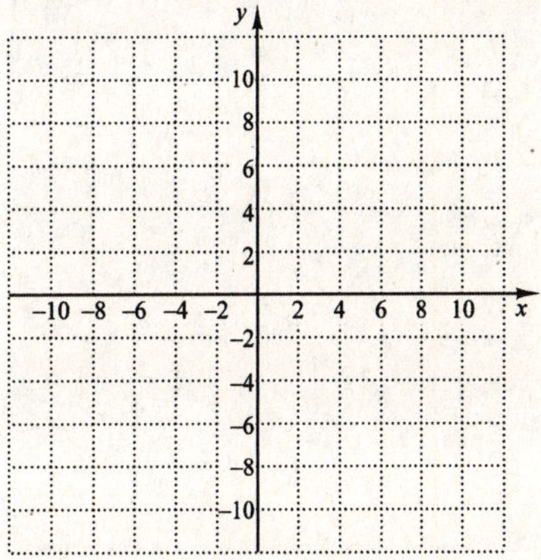

19. $f(x) = 0.4e^x$

19.

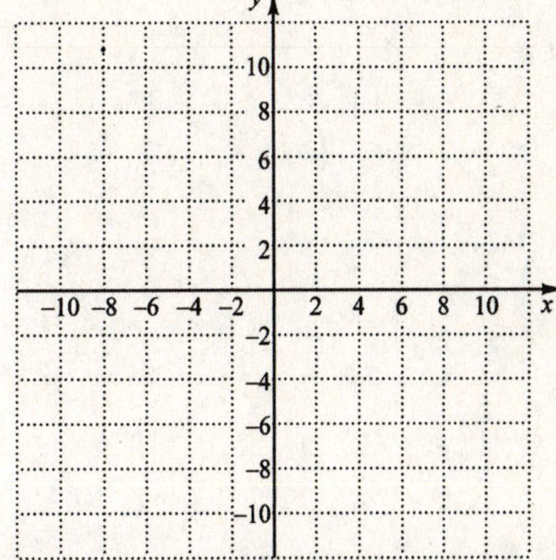

345

20. $g(x) = -\ln x$

20.

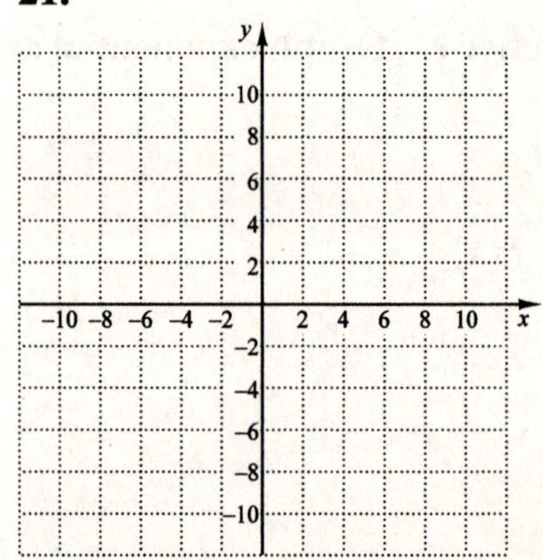

21. $g(x) = \ln(x-2)$

21.

346

Chapter 12 EXPONENTIAL AND LOGARITHMIC FUNCTIONS

12.6 Solving Exponential and Logarithmic Equations

Learning Objective
a   Solve exponential equations.
b   Solve logarithmic equations.

Key Terms
Use the vocabulary terms listed below to complete each statement in Exercises 1–4. Terms may be used more than once.

exponential   **logarithmic**

1. The equation $2^x = 10$ is an example of a(n) _____________________ equation.

2. The equation $\log_2 x = 10$ is an example of a(n) _____________________ equation.

3. The principle of _____________________ equality states that for any real number b, where $b \neq -1, 0,$ or 1, $b^x = b^y$ is equivalent to $x = y$..

4. The principle of _____________________ equality states that for any logarithmic base a, and for $x, y > 0$, $x = y$ is equivalent to $\log_a x = \log_a y$.

Objective a   Solve exponential equations.

Solve.

5. $2^x = 11$ 5. _________________

6. $e^t = 200$ 6. _________________

7. $8 = 2^{x-1}$ 7. _________________

8. $3.6^x = 25$

8. _______________

9. $4e^{3x} = 10$

9. _______________

10. $5^{3x} = 125$

10. _______________

11. $4^{2x-1} = 64$

11. _______________

12. $3^x = 10$

12. _______________

13. $e^{-0.05t} = 0.09$

13. _______________

14. $4^{x+3} = 3^{x-1}$

14. _______________

348

Objective b   Solve logarithmic equations.

Solve.

15. $\log_3 x = -2$ 15. ______________

16. $\log_9 x = \dfrac{1}{2}$ 16. ______________

17. $\ln x = 9$ 17. ______________

18. $\ln(2x + 3) = 6$ 18. ______________

19. $2\log x = 3$ 19. ______________

20. $\log_2(9 - x) = 3$ 20. ______________

21. $\log(x+3) + \log x = 1$

21. _______________

22. $\log x - \log(x+1) = 1$

22. _______________

23. $\log_5(x+2) = 3 + \log_5(x-4)$

23. _______________

24. $\log_2(x+6) + \log_2 x = 4$

24. _______________

350

Chapter 12 EXPONENTIAL AND LOGARITHMIC FUNCTIONS

12.7 Mathematical Modeling with Exponential and Logarithmic Functions

> **Learning Objectives**
> a   Solve applied problems involving logarithmic functions.
> b   Solve applied problems involving exponential functions.

Key Terms
Use the vocabulary terms listed below to complete each statement in Exercises 1–6.

decay	decibel	doubling time
half-life	growth	pH

1. The _________________ scale is used to measure the volume of a sound.

2. The _________________ of a liquid is a measure of its acidity.

3. An exponential _________________ model is a function of the form
   $P(t) = P_0 e^{kt}, k > 0$.

4. The _________________ is the amount of time necessary for a population to double
   in size.

5. An exponential _________________ model is a function of the form
   $P(t) = P_0 e^{-kt}, k > 0$.

6. The _________________ is the amount of time necessary for half of a quantity to
   decay.

Objective a   Solve applied problems involving logarithmic functions.

Solve.

7.   The intensity of the sound from a large orchestra is   7. _______________
      about 6.3×10^{-3} W/m^2. How loud in decibels is this
      sound level? Use

$$L = 10 \cdot \log \frac{I}{I_0}, \text{ where } I_0 = 10^{-12} \text{ W/m}^2.$$

351

8. The sound level in a subway is 88 dB. What is the
intensity of this sound? Use
$$L = 10 \cdot \log \frac{I}{I_0}, \text{ where } I_0 = 10^{-12} \text{ W/m}^2.$$

9. The hydrogen ion concentration of ammonia is
3.16×10^{-11} moles per liter. Find the pH. Use
$$\text{pH} = -\log \left[\text{H}^+ \right].$$

10. The pH of lemon juice is 2.0. What is the hydrogen
ion concentration of lemon juice? Use
$$\text{pH} = -\log \left[\text{H}^+ \right].$$

11. Students in a literature class took a final exam.
They took equivalent forms of the exam at
monthly intervals thereafter. The average score
$S(t)$, in percent, after t months was found to be
given by $S(t) = 72 - 15\log(t+1), t \geq 0$.

a) What was the average test score when they
   initially took the test, $t = 0$?
b) What was the average score after 3 months?
c) What was the average score after 12 months?

352

Name: Date:
Instructor: Section:

Objective b   Solve applied problems involving exponential functions.

Solve.

12. The population of the Marshall Islands, in
    thousands, t years after 2003 can be approximated
    by $P(t) = 56.4(1.029)^t$ (*Source*: Based on data from
    Time Almanac, 2004).
    a) In what year will the population reach 75,000?
    b) Find the doubling time.

12.

a)______________

b)______________

13. A college loan of $45,000 is made at 4% interest,
    compounded annually. After t years, the amount due,
    A, is given by the function $A(t) = 45,000(1.04)^t$.
    a) After what amount of time will the amount due
    reach $50,000?
    b) Find the doubling time.

13.

a)______________

b)______________

14. Suppose that P_0 is invested in a savings account
     where interest is compounded continuously at 3.5%
     per year.
     a) Using the model $P(t) = P_0 e^{kt}$, express $P(t)$ in
     terms of P_0 and 0.035.
     b) Suppose that $2000 is invested. What is the
     balance after 1 yr? After 2 yr?
     c) When will an investment of $2000 double itself?

14.

a)______________

b)______________

c)______________

353

15. In 2006, the population of Uganda was 28 million and the exponential growth rate was 3% per year (*Source*: *Time Almanac*).
   a) Find the exponential growth function.
   b) Predict the population of Uganda in 2010.
   c) When will the population of Uganda reach 40 million?

15.
a)______________

b)______________

c)______________

16. The exponential growth rate of the population of Mexico is 1.16% (*Source*: *The World Factbook*, www.cia.gov). What is the doubling time?

16. ______________

17. How old is an archaeological discovery that has lost 15% of its carbon-14? Use $P(t) = P_0 e^{-0.00012t}$.

17. ______________

18. The decay rate of manganese-52 is 12.5% per day. What is its half-life? Use $P(t) = P_0 e^{-kt}$.

18. ______________

Chapter 1 INTRODUCTION TO REAL NUMBERS AND ALGEBRAIC EXPRESSIONS

Section 1.1

Key Terms

1. algebraic expression
3. constant
5. evaluating

Objective a

7. 240 cm^2
9. $270
11. 30
13. 4

Objective b

15. $x+8$, or $8+x$
17. $b \div x$, or $\dfrac{b}{x}$, or b/x, or $b \cdot \dfrac{1}{x}$
19. pq
21. $7d+10$, or $10+7d$
23. $4y+7$, or $7+4y$
25. $p-0.25p$

Section 1.2

Key Terms

1. set
3. graph
5. natural numbers
7. integers

Objective a

9. $-3;\ 55$

Objective b

11. 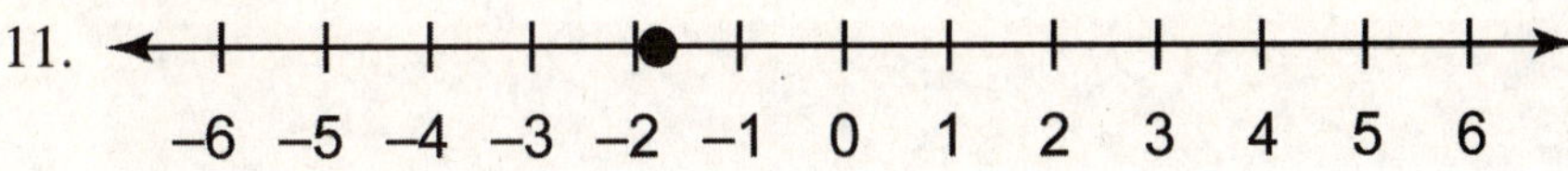

Objective c

13. -0.375
15. $2.1\overline{6}$
17. -1.25

Objective d

19. >

21. >

23. >

25. >

27. False

29. $x \le -3$

Objective e

31. 16

33. $\dfrac{1}{4}$

35. $4\dfrac{7}{8}$

Section 1.3

Key Terms

1. positive

3. zero

Objective a

5. −5

7. 0

9. −13

11. 25

13. −5.1

15. $-\dfrac{5}{6}$

17. $-\dfrac{43}{72}$

19. 75

Objective b

21. 48

23. $-\dfrac{11}{16}$

25. −19

27. −28

Objective c

29. 3-yd gain

356

Section 1.4

Key Terms

1. difference

Objective a

3. −10

5. 15

7. 6

9. 7

11. $-\dfrac{13}{12}$

13. 7.8

15. 7.75

17. 29

19. 17.7

Objective b

21. 130 ft

Section 1.5

Key Terms

1. negative

3. positive

Objective a

5. −20

7. 56

9. 214.4

11. $-\dfrac{1}{6}$

13. 80

15. 0

17. 400; −80

Objective b

19. −12 lb

21. $39.71

Section 1.6

Key Terms

1. not defined

3. negative

Objective a

5. -9

7. 25

9. Not defined

Objective b

11. $\dfrac{9}{4}$

13. $-\dfrac{1}{12}$

15. $-3t$

Objective c

17. $3 \cdot \left(\dfrac{1}{8} \right)$

19. $-6.8 \cdot \left(\dfrac{1}{13} \right)$, or $6.8 \cdot \left(-\dfrac{1}{1.3} \right)$

21. $(2x-1)\left(\dfrac{1}{4} \right)$

23. $-\dfrac{25}{42}$

25. $\dfrac{9}{10}$

27. -4

Objective d

29. 66.1%

Section 1.7

Key Terms

1. equivalent

3. associative

5. terms

7. like

Objective a

9. $\dfrac{3x}{10xy}$

11. $\dfrac{6b}{5}$

Objective b

13. $xy+18$, or $18+yx$

15. $7(ab)$

17. $(9x)y$, $x(9y)$, $9(xy)$; answers may vary

Objective c

19. $5x+35+10n$

21. $\dfrac{3}{4}y-9$

23. $3.22a+2.24b+4.76$

25. $5x,\ 6y,\ -7z$

Objective d

27. $6(2y+5z)$

29. $-5(t-4),\ \text{or } 5(-t+4)$

31. $\dfrac{1}{4}(3x-7y+1)$

Objective e

33. $-3t$

35. $-37c+2$

37. $\dfrac{16}{3}y+\dfrac{2}{5}z-11$

Section 1.8

Key Terms

1. grouping

3. divisions

Objective a

5. $-3x-y+10z$

7. $10m+3n+18$

Objective b

9. $4x-7$

11. $-2p+13q$

13. $-7a+15b$

Objective c

15. -18

17. $7x-1$

19. $-45x-35$

Objective d

21. -17

23. 31

25. 44

27. 63

29. -57

31. 10

33. $-\dfrac{14}{9}$

Chapter 2 SOLVING EQUATIONS AND INEQUALITIES

Section 2.1

Key Terms

1. equation

3. equivalent equations

Objective a

5. No

7. Yes

9. Yes

Objective b

11. 12

13. −11

15. −2

17. −7

19. $\dfrac{59}{40}$

21. −1.5

Section 2.2

Key Terms

1. principle

3. identity

Objective a

5. 7

7. −20

9. −6

11. 3

13. 45

15. $-\dfrac{4}{3}$

17. $\dfrac{4}{3}$

19. 6

21. 135

Section 2.3

Key Terms

1. distributive laws

3. infinitely many solutions

Objective a

5. 8

7. -11

9. -3

Objective b

11. 7

13. -4

15. 2

17. 13

19. $-\dfrac{1}{8}$

21. $\dfrac{49}{3}$

23. $-\dfrac{5}{7}$

Objective c

25. 6

27. 1

29. All real numbers

31. 11

33. No solution

Section 2.4

Key Terms

1. formula

Objective a

3. (a) 195 mi; (b) $t = \dfrac{d}{65}$

5. 126 cm²

Objective b

7. $d = \dfrac{c}{8}$

9. $x = 18 - y$

11. $x = \dfrac{P - b}{a}$

13. $r^2 = \dfrac{A}{4\pi}$

15. $w = \dfrac{4k}{c}$

Section 2.5

Key Terms

1. of

3. what number

Objective a

5. 65%
9. 154
13. 350
17. 64.3%
21. 39 lb

7. 18
11. 16%
15. 25
19. (a) 20%; (b) $36
23. 17.8% decrease

Section 2.6

Key Terms

1. familiarize
5. state

3. solve

Objective a

7. $14.56
11. 64, 66, 68
15. 24°, 48°, 108°
19. $44

9. 0.5 in.
13. $64
17. 9 oz

Section 2.7

Key Terms

1. inequality

3. equivalent

Objective a

5. (a) No; (b) no; (c) yes; (d) yes; (e) no

362

Objective b

7. $m < 2$

9. $-2 \le x < 5$

Objective c

11. $\{x \,|\, x < -1\}$

13. $\{t \,|\, t > 25\}$

15. $\{y \,|\, y \ge 21\}$

Objective d

17. $\{x \,|\, x \le -2\}$

19. $\{x \,|\, x \le -6\}$

21. $\left\{p \,\middle|\, p > -\dfrac{7}{4}\right\}$

Objective e

23. $\{y \,|\, y > 5\}$

25. $\{x \,|\, x \ge 4\}$

27. $\{y \,|\, y > -3\}$

29. $\{y \,|\, y \le 2\}$

Section 2.8

Key Terms

1. d

3. b

5. b

Objective a

7. $n \le 13$

9. $45 \text{ mph} < s < 65 \text{ mph}$

11. $y \le 50$

13. $a \le \$1500$

Objective b

15. $\{x \,|\, x \ge 66\}$

17. 300 or fewer cards

19. Lengths greater than or equal to 6 feet; lengths less than or equal to 6 feet

21. Dates at least four weeks after August 1

Chapter 3 GRAPHS OF LINEAR EQUATIONS

Section 3.1

Key Terms

1. axes

3. coordinates

5. graph

Objective a

7. II

9. On an axis

11. I, III

Objective b

Objective c

13. No

15. Yes

17. $x + 3y = 5$

$$\frac{2 + 3 \cdot 1 \,?\, 5}{5 \quad | \qquad \text{True}}$$

$x + 3y = 5$

$$\frac{5 + 3 \cdot 0 \,?\, 5}{5 \quad | \qquad \text{True}}$$

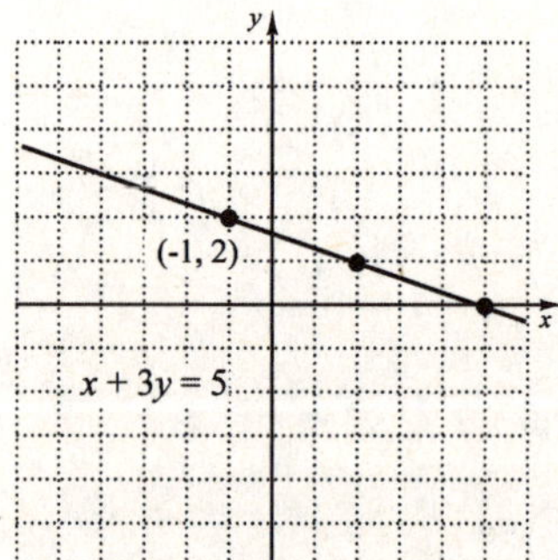

Section 3.2

Objective a

1.

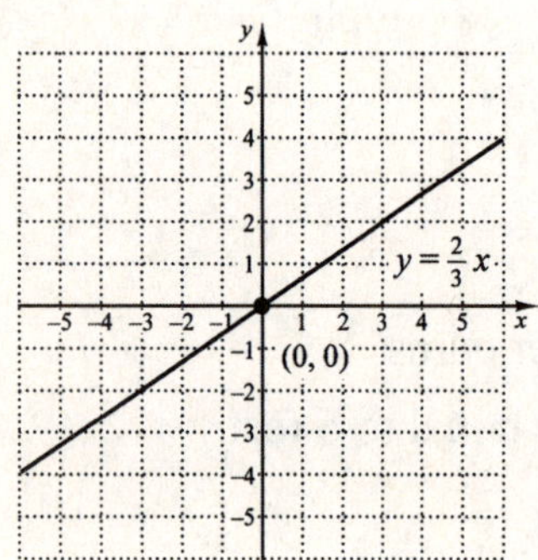

3.

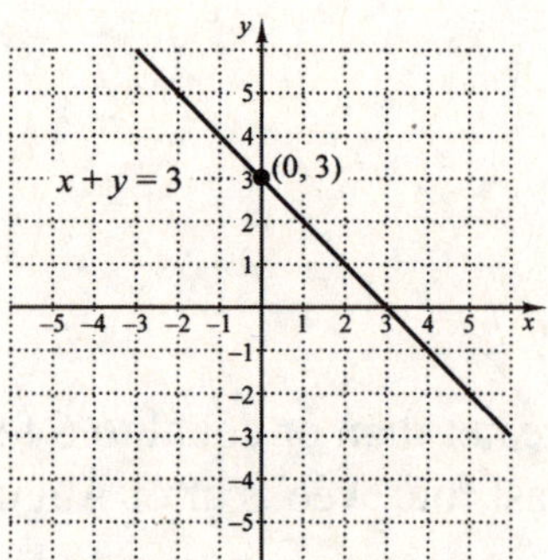

364

5. 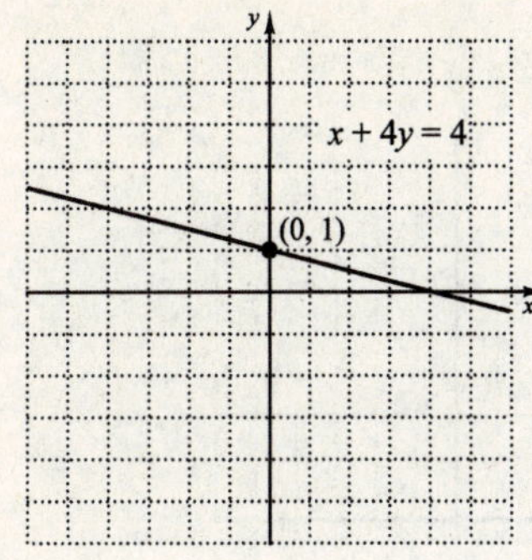

7. (a) \$435, \$477, \$603, \$666; (b) \$540

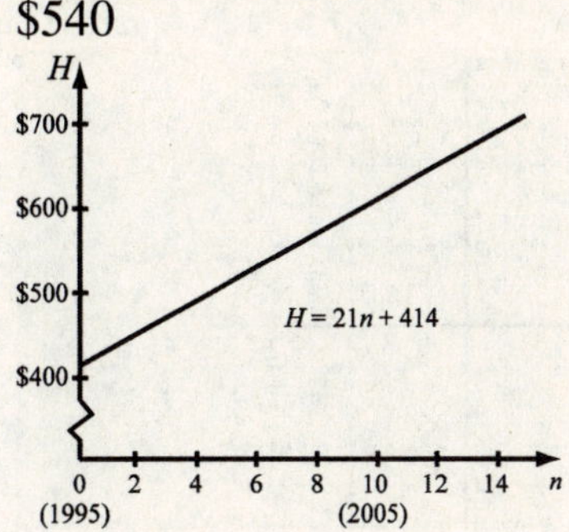

Section 3.3

Key Terms

1. x-intercept

3. vertical line

Objective a

5. (a) $(0,3)$; (b) $(2,0)$

7. (a) $(0,6)$; (b) $(5,0)$

9. (a) $\left(0,\dfrac{5}{2}\right)$; (b) $\left(-\dfrac{5}{3},0\right)$

11. 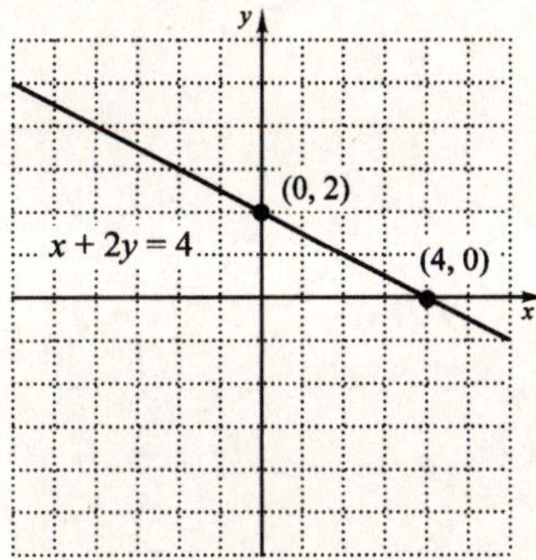

12. 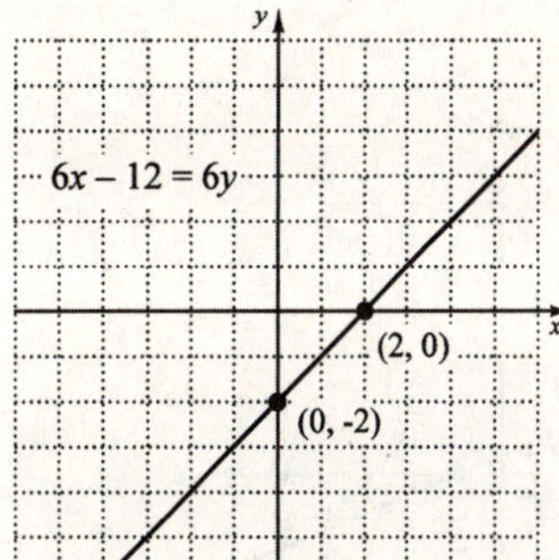

15. 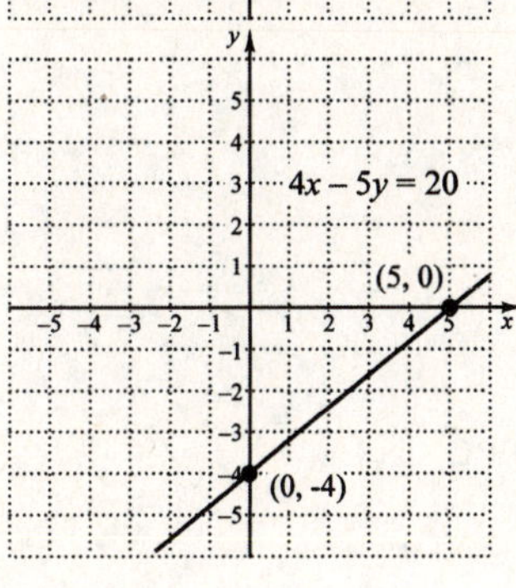

16.

365

Objective b

17. 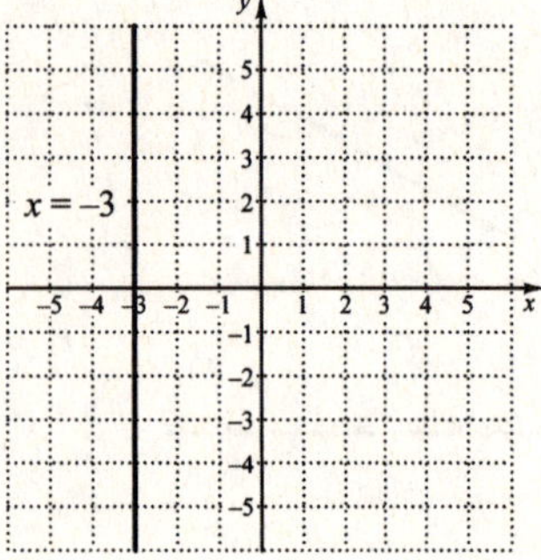

19.

21. $x = -2$

Section 3.4

Key Terms

1. rise

3. slope

Objective a

5. $-\dfrac{4}{9}$

7. Not defined

9. $-\dfrac{2}{5}$

11. $-\dfrac{2}{5}$

13. 0

Objective b

15. -6

17. Not defined

19. $-\dfrac{1}{2}$

21. 13.8

Objective c

23. 2.5%

25. 20 mpg

366

Chapter 4 POLYNOMIALS: OPERATIONS

Section 4.1

Key Terms

1. exponent

3. x-squared

Objective a

5. $7 \cdot 7 \cdot 7$

7. $(8x)(8x)(8x)(8x)$

Objective b

9. 1

11. 1

Objective c

13. 32

15. 13

Objective d

17. 7^{10}

19. x^{21}

Objective e

21. 11^5

23. t^9

Objective f

25. $\dfrac{1}{4^2} = \dfrac{1}{16}$

27. x^5

29. x^{-4}

31. $\dfrac{1}{5^2} = \dfrac{1}{25}$

33. $\dfrac{1}{c}$

35. $\dfrac{1}{7}$

37. $\dfrac{1}{t^4}$

Section 4.2

Key Terms

1. exponential notation

Objectives a,b

3. 3^8

5. x^{14}

7. $\dfrac{1}{x^3 y^6}$

9. $\dfrac{b^6 c^{12}}{a^4}$

11. $\dfrac{x^{10}}{9}$

13. $\dfrac{p^{10} q^{20}}{n^5}$

Objective c

15. 4.6×10^{11}

17. 4.05×10^{-7}

19. 402,000,000

21. 0.0001

Objective d

23. 7.35×10^{-4}

25. 3.36×10^{-10}

Objective e

27. 9.4608×10^{15} m

Section 4.3

Key Terms

1. monomial

3. like terms

5. degree

7. binomial

Objective a

9. $19; -1$

11. $1, -2, -3, -0.75, 6$

Objective b

13. $5, -4x^2, x$

Objective c

15. $4x^3$ and $2x^3$; $-6x^2$ and $-x^2$

368

Objective d

17. $3, \dfrac{1}{4}, -7$

Objective e

19. $-9x$

21. $\dfrac{5}{4}x^4 - x^2 - 11$

Objective f

23. $x^5 + 5x^3 + 2x^2 - x - 6$

25. $9x^2 - 9x + 11$

Objective g

27. $4, 1, 0; 4$

Objective h

29. $x^5, \ x^4, \ x^3, \ x^0$

31. $x^3 + 0x^2 + 0x - 8;$
$\quad x^3 \qquad\qquad - 8$

Objective i

33. monomial

35. none of these

Section 4.4

Key Terms

1. sign

Objective a

3. $-x - 5$

5. $15x^2 - 4$

7. $10 - 6x$

9. $3x^7 + 8x^6 - 17x^5 + 10x^3 + 4x + 3$

Objective b

11. $10y$

13. $-6x^5 + 7x - 10$

15. $5x^2 + 6x - 8$

Objective c

17. $-x^2 - 7x + 2$

19. $8x^4 - 5x^3 + x^2 - 4$

21. $x^3 - \dfrac{9}{5}x$

23. $6x^4 + 16x^3 - x^2 - 7x - 11$

Objective d

25. $12x + 30$

27. $(n+9)(n+8);\ 8n + 72 + n^2 + 9n,\ or\ n^2 + 17n + 72$

29. $x^2 - 16\pi$

Section 4.5

Key Terms

1. term

Objective a

3. $21a^9$

5. $\dfrac{1}{20}x^9$

Objective b

7. $-8x^2 + 12x$

9. $12y^8 + 30y^{11}$

Objective c

11. $x^2 + 9x + 8$

13. $4x^2 - 12x + 9$

15. $x^2 + \dfrac{5}{6}x - 1$

17. $(x+3)(x+7),\ x^2 + 10x + 21$

19.

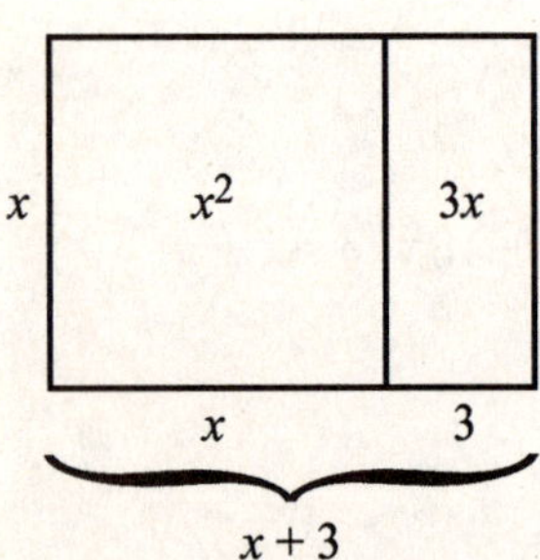

Objective d

21. $x^3 - x^2 - 4$
25. $-12x^5 - 20x^4 + 8x^3 + 15x^2 - 5x$
29. $y^4 + 5y^3 + 2y^2 - 9y - 7$

23. $3x^4 - 2x^3 - 11x^2 + 8x - 4$
27. $6n^4 + 13n^3 - 24n^2 - 22n + 15$

Section 4.6

Key Terms

1. binomials

3. difference

Objective a

5. $x^4 + 2x^3 + 3x + 6$
9. $t^2 - 0.04$
13. $n^{24} - 25$

7. $2a^2 + a - 15$
11. $1 - 3x - 40x^2$

Objective b

15. $x^2 - 9$
19. $x^{10} - 4x^2$

17. $16t^6 - 25$

Objective c

21. $x^2 + 10x + 25$

25. $36 - 24x^5 + 4x^{10}$

23. $t^2 - \dfrac{2}{3}t + \dfrac{1}{9}$

Objective d

27. $25 - 30x^2 + 9x^4$
31. $y^3 + 8$
35. $t^2 + 4t + 4$

29. $36t^4 - \dfrac{1}{25}$
33. $41;\ 81$

371

Section 4.7

Key Terms

1. degree

Objective a

3. -59

5. 1423 calories

Objective b

7. Coefficients: 2, -3, 4, -7; degrees: 3, 2, 4, 0; 4

Objective c

9. $c - 3d$

11. $22xy + 13xz$

Objective d

13. $-x^2 - 2xy - 5y^2$

15. $-u^2 - 4uv - v^2$

Objective e

17. $2x^3 - 3x^3 y + xy^3$

19. $5a^5 b^2 + 6a - 2b + 11$

Objective f

21. $8x^2 - 2xy - 3y^2$

23. $a^2 b^2 + a^2 b^3 + 3b^2 + a^2 + a^2 b + 3$

25. $c^2 + 4cd + 4d^2$

27. $10x^3 - 90xy^2$

29. $a^2 + 2ab + b^2 - 25$

Section 4.8

Key Terms

1. quotient

3. dividend

Objective a

5. $6x^7$

7. $17c^2 d$

9. $13x^4 - 16x^3 + 20x$

11. $-11a^6 + 13a^4 - 16a^2$

13. $4a^2 - 2a + \dfrac{1}{3}$

372

Objective b

15. $x+5$

17. $x+5+\dfrac{8}{x-1}$

19. $x^3-x^2+x-1+\dfrac{2}{x+1}$

21. x^3+5

23. $2x^3+3x^2+2x-1+\dfrac{-2}{2x-3}$

Chapter 5 POLYNOMIALS: FACTORING

Section 5.1

Key Terms

1. factor

3. factorization

Objective a

5. x

7. 3

9. x

Objective b

11. $x(x-10)$

13. $5x^2(2x^3-1)$

15. $11x^2y(x^2y^2-3xy^4-5)$

17. $p^3q^2(q^2+1-pq-p^3q^6)$

19. $0.7x(2x^3-5x^2+6x+10)$

Objective c

21. $(x^3+3)(x+1)$

23. $(x^2+5)(x+2)$

25. $(4p^2+5)(2p-5)$

27. $(3x^2-1)(x-2)$

Section 5.2

Key Terms

1. leading coefficient

3. positive

Objective a

5. $(x+3)(x+6)$

7. $(a-4)(a+5)$

9. $(t-3)(t-10)$

11. Prime

13. $x(x-8)(x+5)$

15. $(t^2+2)(t^2+5)$

17. $(x+11)^2$

19. $x^2(x-15)(x+2)$

21. $(t-25)(t-3)$

23. $-1(t+12)(t-5)$, or $(-t-12)(t-5)$, or $(t+12)(-t+5)$

25. $(m+6n)(m-11n)$

Section 5.3

Key Terms

1. First

3. Inside

Objective a

5. $(2x+3)(x-4)$

7. $(2x-5)(3x-1)$

9. $(4x-3)(4x+5)$

11. $(7x-3)(3x-7)$

13. $-1(2x+7)(x-1)$, or $(-2x-7)(x-1)$, or $(2x+7)(-x+1)$

15. $5(3x+5)(x+2)$

17. $8(2x+1)(x-5)$

19. $p^2(5p+2)(2p-5)$

21. $(9y+2)^2$

23. $(3m+2n)(2m-5n)$

25. $(2p+3q)(7p+q)$

Section 5.4

Key Terms

1. common factor

3. sum

5. grouping

Objective a

7. $(x+3)(x+4)$

9. $(4y+1)(6y-5)$

11. $(3x-2)(x+5)$

13. $(2x+7)(2x-3)$

374

15. $(7x-1)(2x-9)$

17. $-1(2a-1)(a+3),\text{ or }(-2a+1)(a+3),$
$\text{or}(2a-1)(-a-3)$

19. $(5t-9)(3t+1)$

21. $t(2t+3)(3t+5)$

23. $p^2(6p-5)(2p+1)$

25. $(2x^2-5)(4x^2-3)$

27. Prime

29. $2(3s+7t)(s+t)$

Section 5.5

Key Terms

1. trinomial square

3. sum of squares

Objective a

5. No

Objective b

7. $(x-3)^2$

9. $(m^2+5)^2$

11. $x(x-10)^2$

13. $(2m+5n)^2$

15. $(x-4y)^2$

Objective c

17. Yes

19. No

Objective d

21. $(y+10)(y-10)$

23. $(2m+7n)(2m-7n)$

25. $(0.2a+0.05)(0.2a-0.05)$

27. $3(x^2+4)(x+2)(x-2)$

29. $\left(3+\dfrac{1}{4}x\right)\left(3-\dfrac{1}{4}x\right)$

Section 5.6

Key Terms

1. b

3. c

Objective a

5. $(4t-1)(16t^2+4t+1)$

7. $(z+w)(z^2-zw+w^2)$

9. $(d-0.2)(d^2+0.2d+0.04)$

11. $\left(x-\frac{1}{3}\right)\left(x^2+\frac{1}{3}x+\frac{1}{9}\right)$

13. $10(5t^2+s^2)(25t^4-5t^2s^2+s^4)$

15. $(x+1)(x^2-x+1)(x-1)(x^2+x+1)$

17. $x^6(x^3+y^4)(x^6-x^3y^4+y^8)$

19. $(r+2)(r^2-2r+4)(r-2)(r^2+2r+4)$

Section 5.7

Key Terms

1. difference

3. grouping

Objective a

5. $5(t+2)(t-2)$

7. $(3x+2)(2x-5)$

9. $x(x-4)(x^2+5)$

11. Prime

13. $2x^2(x+3)^2$

15. $-5(2x-1)(x+5)$, or $5(-2x+1)(x+5)$, or $5(2x-1)(-x-5)$

17. $3pq(p^2+4q^2)$

19. $(a+b)(a-1)$

21. $(3c-2d)^2$, or $(2d-3c)^2$

23. $(m+6n)(m-10n)$

25. $x^3(xy-4)(xy+8)$

27. $(c^4d^2+9)(c^2d+3)(c^2d-3)$

Section 5.8

Key Terms

1. quadratic equation

3. zero

Objective a

5. $-3,\ 10$

7. $0,\ -15$

9. $-\dfrac{3}{4},\ 5$

11. $\dfrac{1}{6},\ \dfrac{1}{12}$

Objective b

13. $-4,\ -1$

15. $0,\ 10$

17. $-5,\ 5$

19. -2

21. $-2,\ \dfrac{1}{3}$

23. $-5,\ -3$

25. $-2,\ 4$

27. $-1,\ 2$

Section 5.9

Key Terms

1. right

3. legs

Objective a

5. Length: 8 ft; width: 3 ft

7. Height 6 cm; base: 14 cm;

9. 132 games

11. 435 handshakes

13. 12 people

15. 16 and 18; -18 and -16

17. Hypotenuse: 25 ft; leg: 20 ft

19. 5000 ft

21. 3 sec

Chapter 6 RATIONAL EXPRESSIONS AND EQUATIONS

Section 6.1

Key Terms

1. rational

3. multiplying

Objective a

5. 0

7. $\dfrac{5}{4}$

9. $-4,\ 4$

Objective b

11. $\dfrac{\left(5a^{2}\right)\left(6c^{2}\right)}{\left(5a^{2}\right)\left(7d^{4}\right)}$

377

Objective c

13. $\dfrac{x^4}{5}$

15. $\dfrac{m-5}{m}$

17. $\dfrac{t+1}{t-9}$

19. $\dfrac{x^2+25}{x+5}$

21. $\dfrac{x+2}{3(x-3)}$

23. -1

Objective d

25. $\dfrac{15x^4}{2}$

27. $\dfrac{x+1}{x-1}$

29. $\dfrac{2x^2}{x+3}$

31. $\dfrac{m-7}{m+5}$

Section 6.2

Key Terms

1. interchange

Objective a

3. $\dfrac{y}{3}$

5. $n+3$

Objective b

7. $\dfrac{5}{4}$

9. $\dfrac{1}{y^2 x^3}$

11. $\dfrac{(t-2)^2}{t^2}$

13. $\dfrac{8}{3}$

15. $\dfrac{5(a-7)}{a-3}$

17. $\dfrac{5c(c-d)}{4(c+d)}$

19. $\dfrac{(2x+1)^2}{(x-7)(3x+1)}$

21. $\dfrac{(t+1)(t+2)^2}{9(t+3)}$

Section 6.3

Key Terms

1. least common multiple

Objective a

3. 120

5. 60

Objective b

7. $\dfrac{11}{20}$

9. $\dfrac{53}{72}$

Objective c

11. $30x^5$

13. $45(t+7)$

15. $(a+3)(a-3)(a+1)$

17. $(x-5)^2(x+5)$

19. $(4+5x)(4-5x)$

21. $12y^3(y-3)^2(y+5)$

Section 6.4

Key Terms

1. least common multiple

3. numerators

Objective a

5. $\dfrac{2}{5}$

7. $\dfrac{4a+7}{a^2}$

9. $\dfrac{8d+5c}{c^2d^2}$

11. $\dfrac{44t+27}{7t(t+3)}$

13. $\dfrac{5y-7}{(y-2)^2}$

15. $\dfrac{7x-41}{(x-7)(x-3)(x+1)}$

17. $\dfrac{2(4x-5)}{(x-2)(x+2)}$

19. $\dfrac{-3x+2}{x-10}$

21. $\dfrac{7x+8}{x-3}$

23. $\dfrac{a^2+10a+11}{(a-7)(a+7)}$

379

Section 6.5

Key Terms

1. denominator

3. numerators

Objective a

5. $\dfrac{6}{x}$

7. $\dfrac{1}{x-1}$

9. $\dfrac{-25r^2 + 23rt + t^2}{5r^2t^2}$

11. $\dfrac{x-8}{(x-5)(x-4)(x+5)}$

13. $\dfrac{12}{x}$

15. $\dfrac{2}{(x+4)(x-4)}$

17. $\dfrac{-3y+5}{4y(y-1)}$

Objective a

19. $\dfrac{23t-7}{t-2}$

21. $\dfrac{x+17}{3x-1}$

23. $\dfrac{-x-1}{4x-1}$

Section 6.6

Objective a

1. $\dfrac{19}{4}$

3. $-\dfrac{13}{10}$

5. $\dfrac{3t+1}{t}$

7. $x-10$

9. $-\dfrac{1}{x}$

11. $\dfrac{x-5}{x-3}$

13. $\dfrac{-7(n^2-3)(n^2+3)}{10n^2(n^2+9)}$

15. 1

380

Section 6.7

Key Terms

1. rational

3. LCM

Objective a

5. $-\dfrac{14}{5}$

7. $\dfrac{37}{2}$

9. $\dfrac{40}{13}$

11. $-7,7$

13. $\dfrac{16}{5}$

15. -8

17. -3

19. $\dfrac{27}{2}$

21. $-\dfrac{35}{4}$

23. $\dfrac{1}{2}$

25. No solution

27. $-1,\dfrac{2}{3}$

Section 6.8

Key Terms

1. ratio

3. proportion

Objective a

5. $1\dfrac{5}{7}$ hr

7. $40\dfrac{10}{11}$ min

9. Maya: 65 km/h; Tara 45 km/h

11. Scooter: 42 mph; motorcycle: 58 mph

Objective b

13. 24 mpg

15. 1000 beans

17. 1464 steps

19. 45 trout

21. $\dfrac{12}{5}$

23. $\dfrac{21}{2}$

Section 6.9

Key Terms

1. direct 3. inverse

Objective a

5. 11; $y = 11x$

Objective b

7. $16\frac{2}{3}$ cm 9. 25 min

Objective c

11. 6; $y = \dfrac{6}{x}$

Objective d

13. 120 m

Objective e

15. $y = \dfrac{18}{x^2}$ 17. $y = \dfrac{11}{10}xz^2$

Objective f

19. 3.125 W/m^2

Chapter 7 GRAPHS, FUNCTIONS, AND APPLICATIONS

Section 7.1

Key Terms

1. input 3. outputs
5. relation

Objective a

7. No

382

Objective b

9. 29; −13; 46.5

11. 0; 6; 6

Objective c

13. 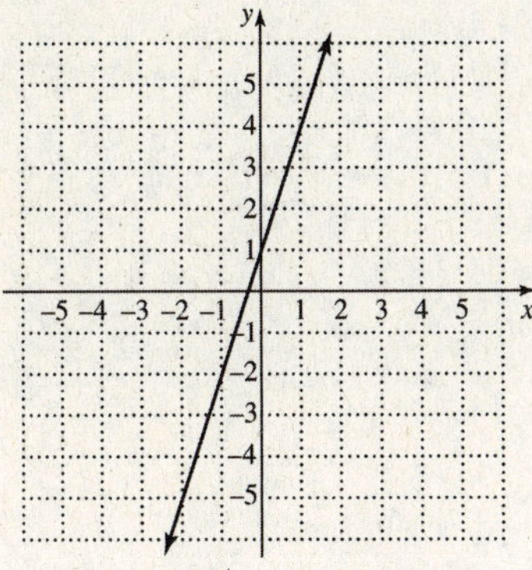

15. 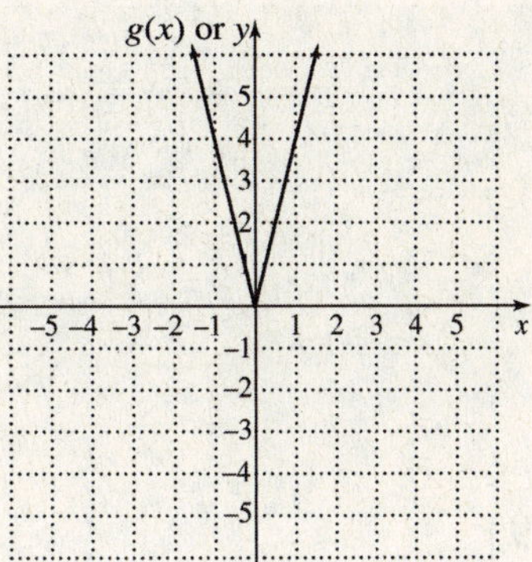

17. 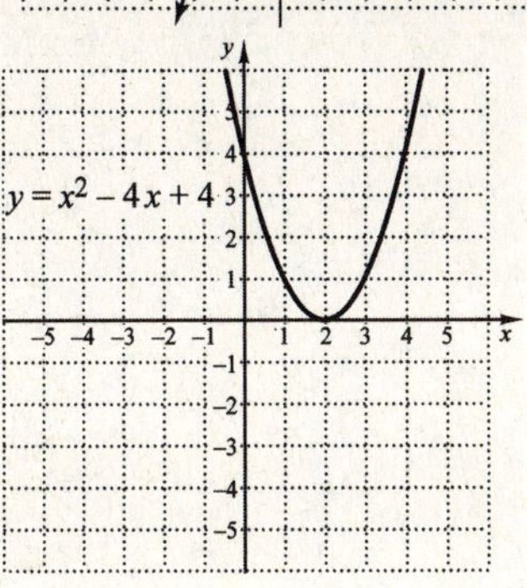

Objective d

19. Yes

21. Yes

Objective e

23. 16 words

Section 7.2

Key Terms

1. domain

3. relation

Objective a

5. a) -3; b) $\left\{x\,\middle|\,-2 \leq x \leq 5\right\}$; c) -1; d) $\left\{x\,\middle|\,-3 \leq x \leq 2\right\}$

7. a) -3; b) $\left\{-5,-4,-3,-2,-1,0,1,2,3,4,5\right\}$; c) -1; d) $\left\{-5,-4,-3,-2,-1,0,1,2,3,4,5\right\}$

9. $\mathbb{R}$

383

11. $\left\{ x \,\middle|\, x \text{ is a real number } and\ x \neq \dfrac{5}{7} \right\}$ or $\left(-\infty, \dfrac{5}{7} \right) \cup \left(\dfrac{5}{7}, \infty \right)$

Section 7.3

Key Terms

1. *m*
3. up
5. rise
7. grade

Objective a

9. $\dfrac{4}{5}; \left(0, -\dfrac{9}{5} \right)$

11. $-\dfrac{2}{3}; (0,3)$

13. $\dfrac{5}{3}; (0,3)$

Objective b

15. $\dfrac{2}{3}$

17. 3

19. $\dfrac{23}{35}$

Objective c

21. 6%

23. $-\dfrac{2}{3}$ gallon per year

Section 7.4

Key Terms

1. parallel
3. vertical
5. *x*-intercept
7. 0

Objective a

9. 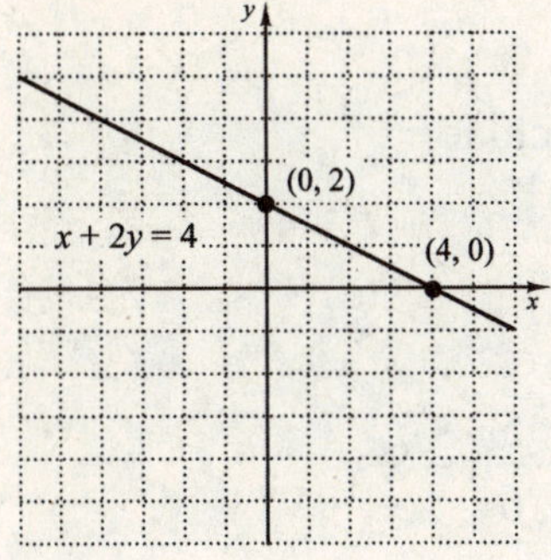

11. 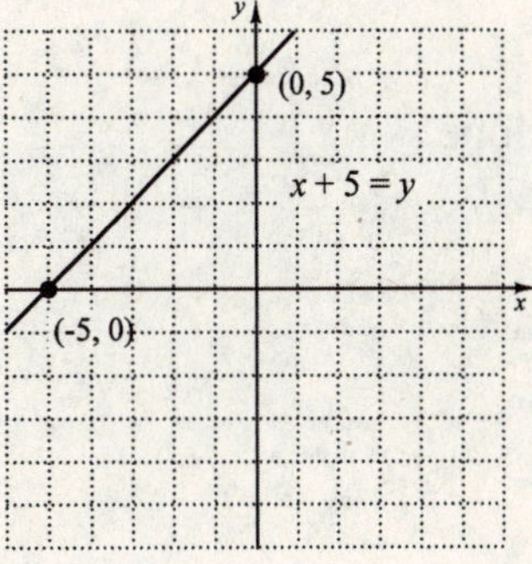

Objective b

13. 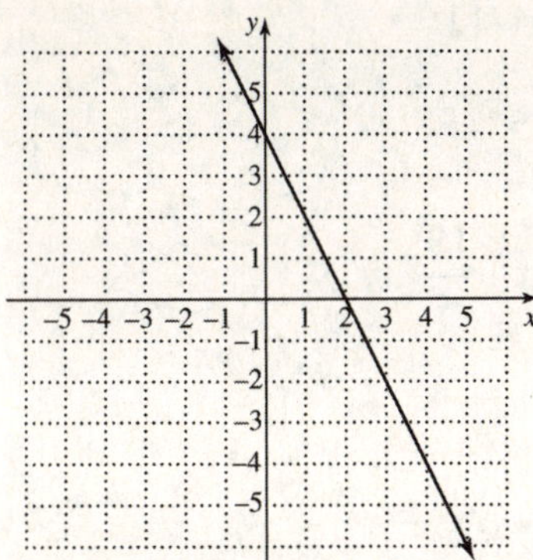

Objective c

15. Not defined

17. $m = 0$

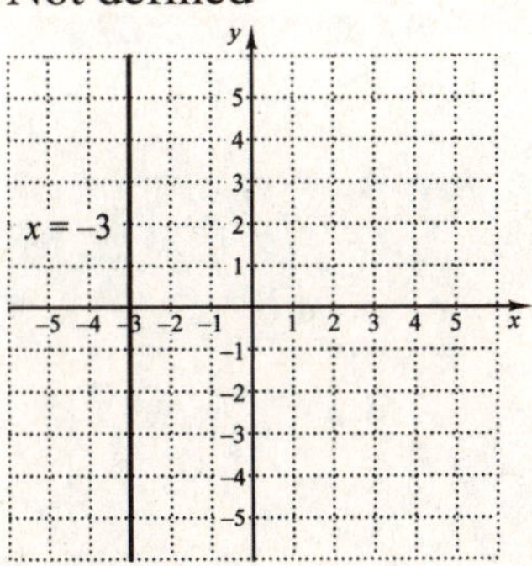

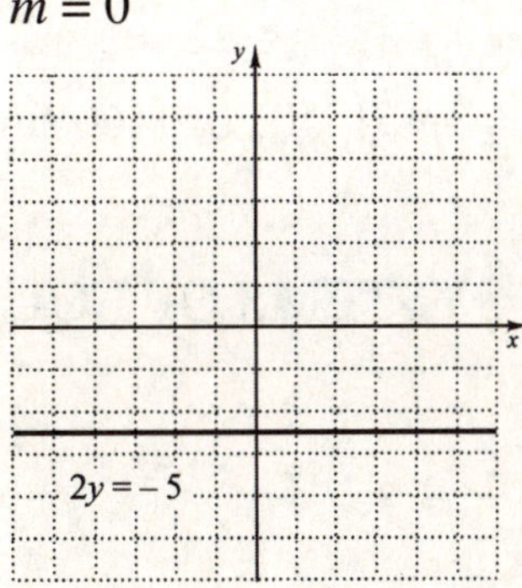

Objective d

19. Yes

21. No

23. No

25. Yes

385

Section 7.5

Key Terms

1. point-slope
3. perpendicular
5. slope

Objective a

7. $y = -15x + 12$

9. $f(x) = -\dfrac{5}{6}x - 8$

Objective b

11. $y = -3x - 12$

13. $y = 2x - 11$

Objective c

15. $y = -x + 11$

17. $y = \dfrac{4}{5}x + \dfrac{19}{5}$

Objective d

19. $y = 4x - 9$

21. $y = 8x + 4$

23. $y = \dfrac{7}{6}x + 8$

Objective e

25. a) $c(t) = \frac{35}{3}t + 1250$; b) 1460 students; c) 2022

Chapter 8 SYSTEMS OF EQUATIONS

Section 8.1

Key Terms

1. system
3. consistent
5. dependent

Objective a

7. $(3, 2)$; consistent; independent

9. $(-2,-1)$; consistent; independent

11. Infinitely many solutions; consistent; dependent

13. $\left(\dfrac{1}{2},-1\right)$; consistent; independent

Section 8.2

Key Terms

1. algebraic

3. solve

Objective a

5. $(2,3)$

7. $(1,1)$

9. $(-7,1)$

11. $\left(\frac{1}{2}, -\frac{3}{4}\right)$

13. $(6,11)$

15. $(0,-2)$

Objective b

17. Length: 36 ft; width: 26 ft

Section 8.3

Key Terms

1. elimination

3. false

Objective a

5. $(7,4)$

7. $(-3,-12)$

9. No solution

11. $\left(\frac{1}{3}, -2\right)$

Objective b

13. Two pointers: 32; three pointers: 9

15. $20°, 70°$

Section 8.4

Key Terms

1. motion

3. rate

Objective a

5. Adults: 250; youth: 325

7. Soup: $2.60; tortilla wrap: $3.80

9. Tropical Punch: 12 L; Caribbean Spring: 6 L

Objective b

11. 3 hr

13. 20 mph

15. In $\dfrac{5}{6}$ hr, or in 50 min

Section 8.5

Objective a

1. $(3,0,1)$

3. $(5,1,-1)$

5. $\left(\dfrac{1}{2},-\dfrac{1}{2},\dfrac{1}{3}\right)$

7. $(2,-3,8)$

Section 8.6

Objective a

1. 36, 19, 23
3. Grapes: 60 calories; melon: 24 calories; pear: 40 calories
5. One-dip cones: 84; two-dip cones: 52; three-dip cones: 64
7. Schools: $5.6 million; roads: $1.2 million; law enforcement: $0.8 million

Chapter 9 MORE ON INEQUALITIES

Section 9.1

Key Terms

1. inequality

3. equivalent

Objective a

5. (a) No; (b) no; (c) yes; (d) yes; (e) no

Objective b

7. $(-3, 6)$

9. $(-\infty,-4)$

388

Objective c

11. $\{x|x \le 3\}$, or $(-\infty, 3]$

13. $\{x|x \ge 6\}$, or $[6, \infty)$

15. $\{z|z > 0\}$, or $(0, \infty)$

17. $\left\{x\middle|x > \dfrac{2}{21}\right\}$, or $\left(\dfrac{2}{21}, \infty\right)$

19. $\{x|x \le 5\}$, or $(-\infty, 5]$

21. $\{x|x \le 11\}$, or $(-\infty, 11]$

23. $\{x|x \le 6\}$, or $(-\infty, 6]$

25. $\left\{y\middle|y \le -\dfrac{69}{8}\right\}$, or $\left(-\infty, -\dfrac{69}{8}\right]$

Objective d

27. $2.99

29. $\{S|S > \$5100\}$

Section 9.2

Key Terms

1. compound inequalities

3. intersection

5. union

7. $\{5, 7, 8, 9, 10, 11, 13\}$

9. $\{f, g\}$

11. $\{2, 3, 5, 10\}$

Objective a

13. $(-2, 6]$

15. $\{x|1 \le x < 8\}$, or $[1, 8)$

17. $\{x|-13 \le x \le 19\}$, or $[-13, 19]$

Objective b

19. $(-\infty, 0) \cup (5, \infty)$

21. $[-3, \infty)$

23. $\{a|a>3\}$, or $(3,\infty)$

25. $\{m|m\leq-3\ or\ m>-2\}$, or $(-\infty,-3]\cup(-2,\infty)$

Objective c

27. $\{w|144.1\ \text{lb}\leq w\leq194.0\ \text{lb}\}$

Section 9.3

Key Terms

1. p

3. $-p$

5. $-p,\ p$

Objective a

7. $5x^2$

9. $19|z|$

11. $2|a|$

13. $9x^4$

Objective b

15. 83

17. $\dfrac{4}{3}$

Objective c

19. $\{0\}$

21. $\left\{-\dfrac{3}{4},\dfrac{13}{4}\right\}$

23. $\varnothing$

25. $\left\{-\dfrac{16}{5},\dfrac{16}{5}\right\}$

Objective d

27. $\left\{-7,\dfrac{5}{3}\right\}$

29. $\left\{-\dfrac{11}{3},\dfrac{3}{7}\right\}$

Objective e

31. $\{x|x\leq-3\ or\ x\geq3\}$, or $(-\infty,-3]\cup[3,\infty)$

33. $\{x|-10<x<4\}$, or $(-10,4)$

390

35. $\left\{x \middle| x \le -\dfrac{9}{2} \ \ or \ \ x \ge \dfrac{7}{2}\right\}$, or $\left(-\infty, -\dfrac{9}{2}\right] \cup \left[\dfrac{7}{2}, \infty\right)$

37. $\left\{x \middle| x \le -\dfrac{1}{4} \ \ or \ \ x \ge \dfrac{29}{4}\right\}$, or $\left(-\infty, -\dfrac{1}{4}\right] \cup \left[\dfrac{29}{4}, \infty\right)$

Section 9.4

Key Terms

1. linear inequality

3. test point

Objective a

5. Yes

Objective b

7. 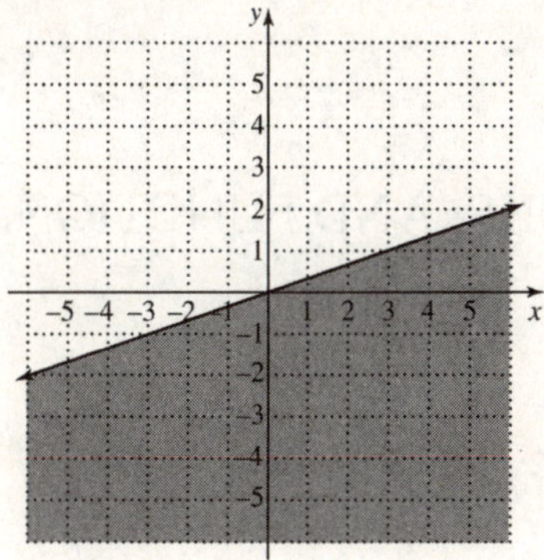

9. 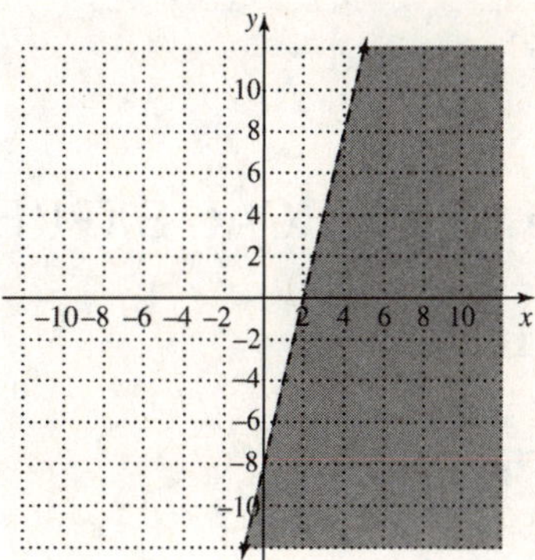

Objective c

11. 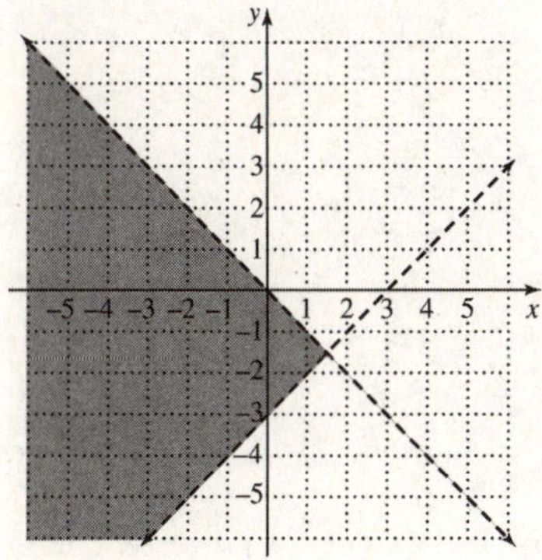

13. 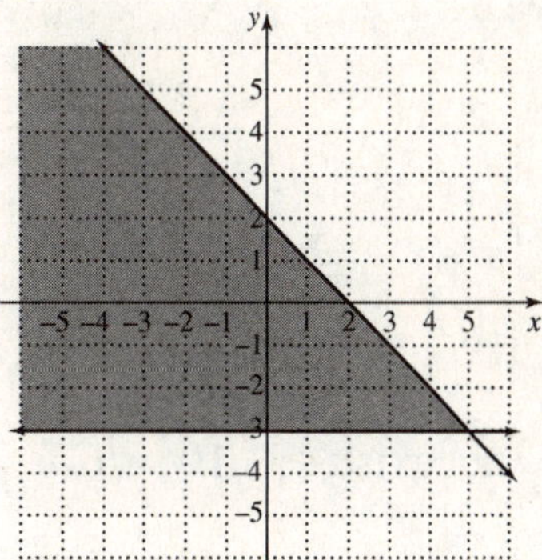

15.

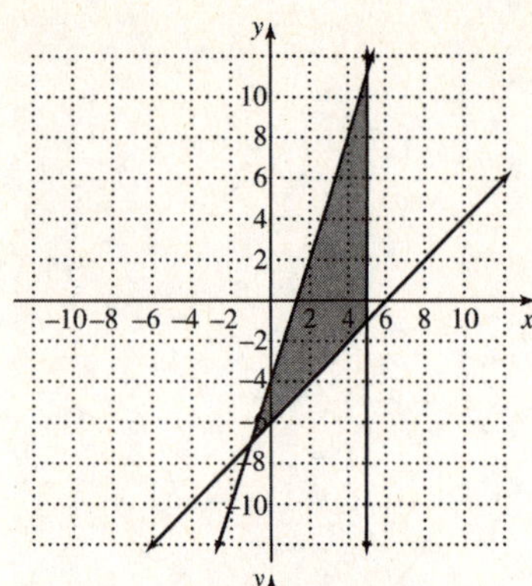

17.

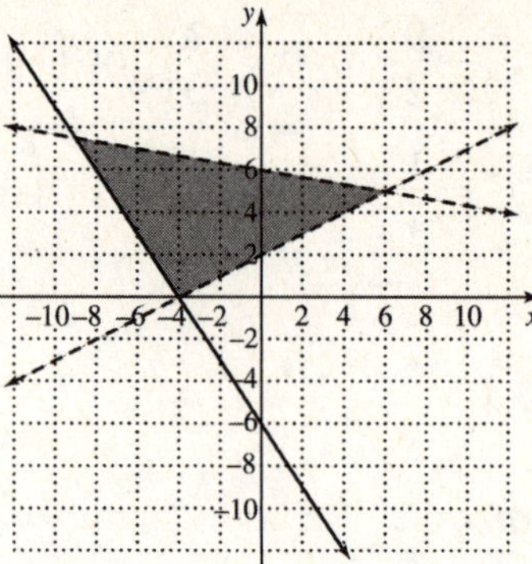

19.

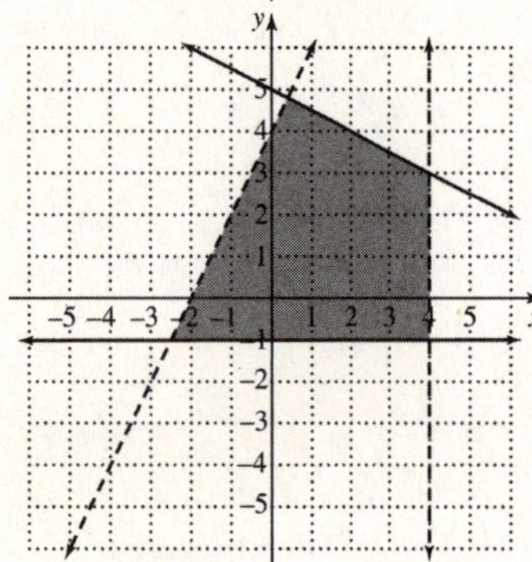

Chapter 10 RADICAL EXPRESSIONS, EQUATIONS, AND FUNCTIONS

Section 10.1

Key Terms

1. square

3. irrational

5. cube

Objective a

7. $8, -8$

9. $-\dfrac{15}{13}$

11. 0.3

13. 23.707

15. -27.619

17. $\dfrac{2s}{3t}$

19. Does not exist; 1; $\sqrt{10}$; $\sqrt{6}$

392

21.

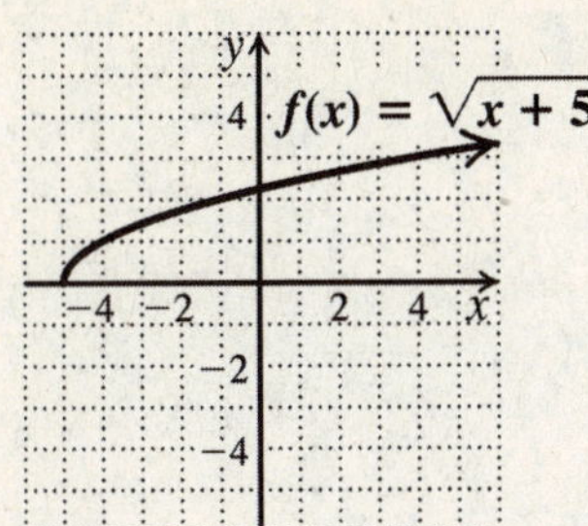

23.

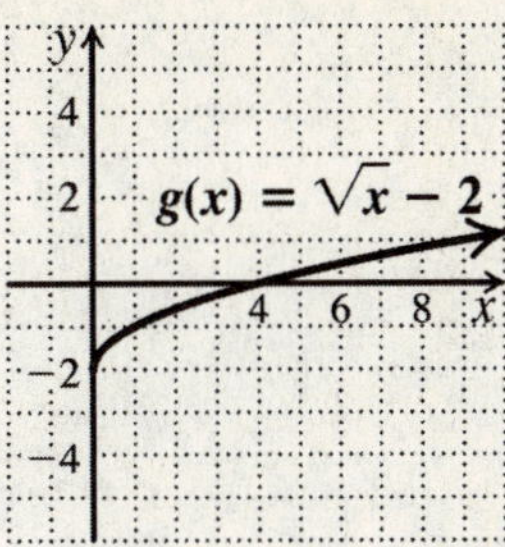

Objective b

25. $|y - 5|$

27. $|2x - 7|$

Objective c

29. $-5t$

31. Does not exist; 1; does not exist; 2

Objective d

33. $-\dfrac{2}{3}$

35. $3ab$

Section 10.2

Key Terms

1. c

3. d

5. f

Objective a

7. $\sqrt[4]{a}$

9. $\sqrt{pqr}$

11. 32

13. $10^{1/5}$

15. $\left(ab^3\right)^{1/4}$

Objective b

17. $\dfrac{1}{(3x)^{1/5}}$

19. $3x^2 y^{4/5}$

Objective c

21. $3^{1/2}$

23. $x^{6/5}$

Objective d

25. x^5

27. $\sqrt[10]{m}$

29. $\sqrt[6]{7776}$

31. $a^9 b^6$

Section 10.3

Key Terms

1. radicands

3. square

Objective a

5. $10\sqrt{2}$

7. $3x^2 \sqrt[3]{y^2}$

9. $x^2 y^5 \sqrt{x}$

11. $-2cd^3 \sqrt[5]{10c^3 d^2}$

13. $\sqrt{110}$

15. $7\sqrt{6}$

17. $28x^5$

19. $cd^3 \sqrt[3]{c^2 d}$

Objective b

21. $\sqrt{6}$

23. $a^2 \sqrt{3b}$

25. $\dfrac{9}{7}$

27. $\dfrac{4p^2 \sqrt{p}}{q^5}$

29. $\dfrac{a^2 b}{c^3} \sqrt[4]{\dfrac{b^3}{c^2}}$

Section 10.4

Key Terms

1. like radicals

Objective a

3. $-3\sqrt{11}$

5. $3\sqrt{3} + 10\sqrt[3]{5}$

7. $\left(2a^2 + 9a\right)\sqrt[3]{5a}$

9. $(2-z)\sqrt{4z-1}$

394

Objective b

11. $6 + 10\sqrt[3]{5}$

13. $-6 - 3\sqrt{10}$

15. $2x + 7 - 2\sqrt{14x}$

17. $\sqrt[4]{72} + \sqrt[4]{225} - \sqrt[4]{24} - \sqrt[4]{75}$

Section 10.5

Objective a

1. $\dfrac{3\sqrt{14}}{10}$

3. $\dfrac{\sqrt[3]{10x^2 y}}{xy}$

5. $\dfrac{\sqrt{21}}{7}$

7. $\dfrac{\sqrt[3]{245}}{7}$

Objective b

9. $\dfrac{5 - \sqrt{3}}{11}$

11. $\dfrac{p + \sqrt{pq}}{p - q}$

Section 10.6

Key Terms

1. radical equation

Objective a

3. $\dfrac{32}{3}$

5. 17

7. 61

9. 0, 49

11. -1000

Objective b

13. 2

15. 3, 7

17. $\dfrac{3}{4}, 1$

Objective c

19. 592 ft

21. 180 ft

23. About 3.6 ft

395

Section 10.7

Objective a

1. $\sqrt{65}$, 8.062

3. $\sqrt{65}$, 8.062

5. $\sqrt{22}$, 4.690

7. $\sqrt{180}$ ft, 13.416 ft

9. 32 in.

Section 10.8

Key Terms

1. i

3. complex

Objective a

5. $7i$

7. $10i\sqrt{2}$, or $10\sqrt{2}i$

9. $8 - 5i\sqrt{6}$, or $8 - 5\sqrt{6}i$

Objective b

11. $12 + 3i$

13. $-3 - 6i$

Objective c

15. -88

17. $8 + 10i$

19. $-28 + 3i$

21. $-33 - 56i$

Objective d

23. 1

25. $-8i$

Objective e

27. $\dfrac{6}{5} - \dfrac{3}{5}i$

29. $-\dfrac{5}{2} - 3i$

Objective f

31. Yes

33. No

Chapter 11 QUADRATIC EQUATIONS AND FUNCTIONS

Section 11.1

Key Terms

1. quadratic

3. complete

Objective a

5. a) $\dfrac{9}{2}i, -\dfrac{9}{2}i,$ or $\pm\dfrac{9}{2}i$; b) no x-intercepts

7. $\pm\frac{9}{2}i$

9. $5\pm\sqrt{2}$; 3.586, 6.414

11. $-\dfrac{2}{3}\pm\dfrac{2\sqrt{2}}{3}$; $-1.609, 0.276$

Objective b

13. 2, 4

15. $-3\pm\sqrt{11}$

17. $\left(2-\sqrt{2},0\right),\left(2+\sqrt{2},0\right)$

19. $\left(\dfrac{1}{4}-\dfrac{\sqrt{5}}{4},0\right),\left(\dfrac{1}{4}+\dfrac{\sqrt{5}}{4},0\right)$

21. $-\dfrac{1}{3}\pm\dfrac{\sqrt{13}}{3}$

23. $-4\pm 3i$

Objective c

25. About 7.8 sec

Section 11.2

Key Terms

1. $\pm\sqrt{k}$

Objective a

3. $-\dfrac{3}{2}\pm\dfrac{\sqrt{17}}{2}$

5. $-\dfrac{1}{2}\pm\dfrac{\sqrt{11}}{2}i$

7. $-2\pm\dfrac{3\sqrt{2}}{2}$

9. $-\dfrac{1}{4}\pm\dfrac{\sqrt{105}}{20}$

11. 1, 3

13. $3\pm 4i$

15. $-1\pm\sqrt{6}$; $-3.449, 1.449$

17. $\dfrac{-2\pm 3\sqrt{2}}{2}$; $-3.121, 1.121$

Section 11.3

Objective a

1. Length: 9 ft; width: 4 ft
3. Base: $2\sqrt{37} - 2$ ft; height: $2 + 2\sqrt{37}$ ft
5. 76 and 77
7. First part: 50 mph; second part: 45 mph
9. Tim: 300 mph, David: 375 mph; or Tim: 125 mph, David 200 mph
11. About 14.4 mph
13. $x = \dfrac{-5 + \sqrt{25 + 12y}}{6}$
15. $q = \dfrac{25p}{M^2}$
17. $v = \dfrac{-4x_0 + 2\sqrt{4x_0^2 + 2at}}{a}$
19. $d = \dfrac{H^2}{2.56}$

Section 11.4

Key Terms

1. discriminant
3. conjugates

Objective a

5. Two real
7. Two real
9. Two nonreal
11. Two real

Objective b

13. $x^2 - 8x + 16 = 0$
15. $5x^2 - 17x - 12 = 0$
17. $x^2 - 90 = 0$

Objective c

19. $\pm 1, \ \pm 4$
21. 25
23. $-\dfrac{1}{6}, \dfrac{1}{5}$
25. $-1, 1024$
27. $(-2, 0), (-1, 0), (3, 0), (4, 0)$

Section 11.5

Key Terms

1. parabola
3. vertex

Objective a, b

5. Vertex: $(-2, 0)$;
    axis of symmetry: $x = -2$

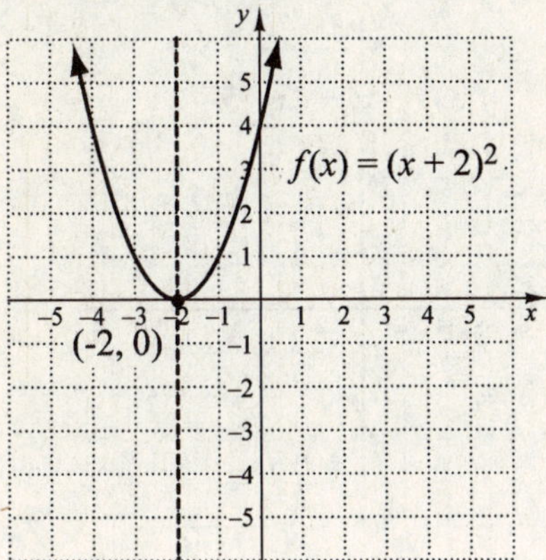

7. Vertex: $(3, 0)$;
    axis of symmetry: $x = 3$

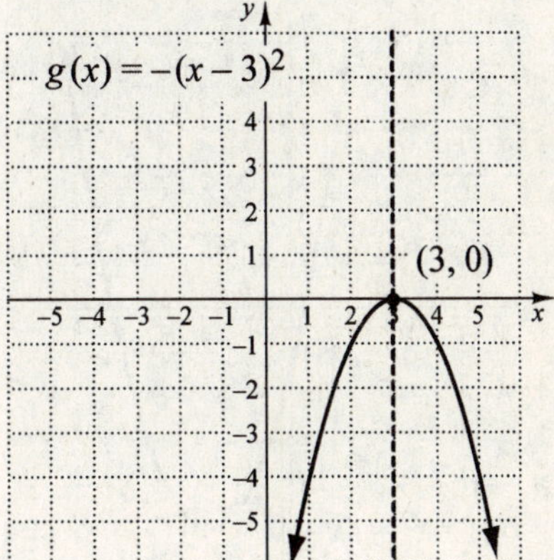

Objective c

9. Vertex: $(-2, -3)$;
    axis of symmetry: $x = -2$;
    minimum: -3

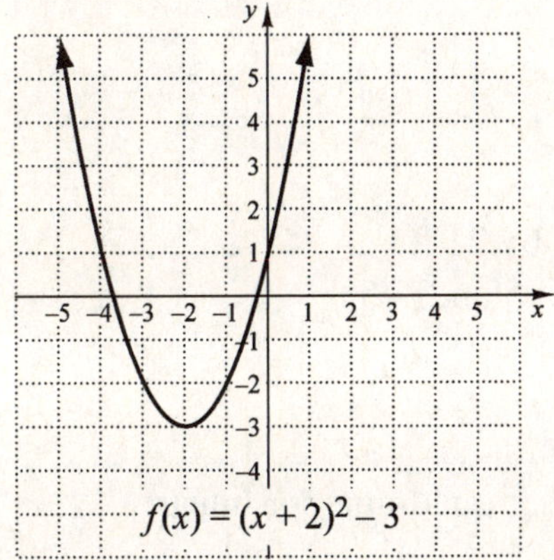

11. Vertex: $(3, 4)$;
    axis of symmetry: $x = 3$;
    maximum: 4

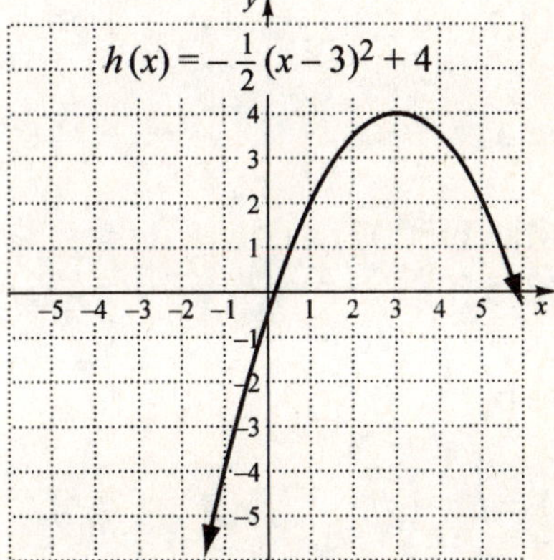

Section 11.6

Key Terms

1. minimum

3. y-intercept

Objective a

5. a) $f(x) = (x + 4)^2 - 21$; b) vertex: $(-4, -21)$; axis of symmetry: $x = -4$

399

7. a) Vertex: $(2,-5)$;
    axis of symmetry: $x=2$;
    b) minimum: -5
    c)

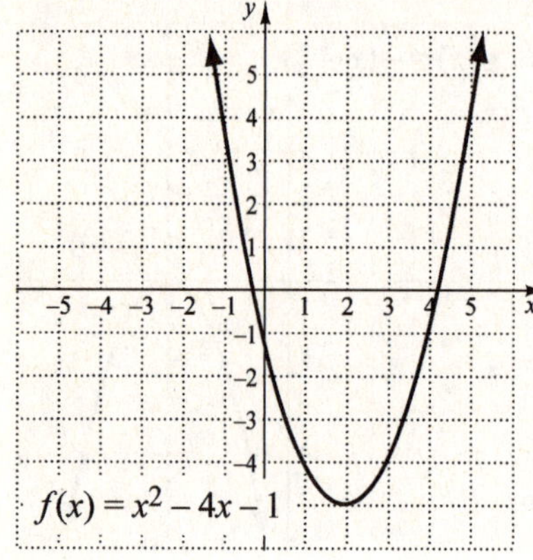

9. a) Vertex: $(1,-4)$;
    axis of symmetry: $x=1$;
    b) maximum: -4
    c)

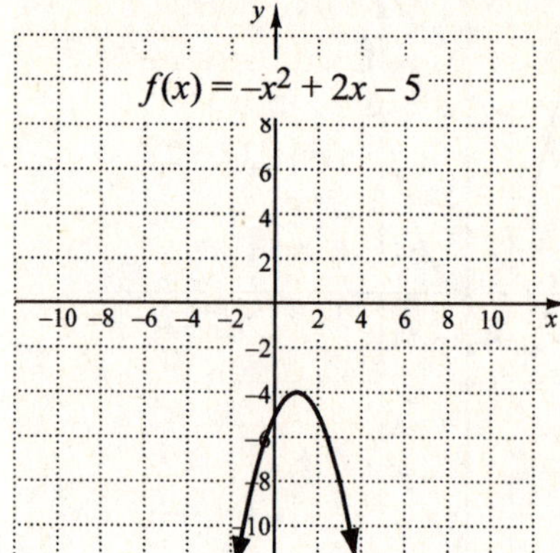

Objective b

11. $(-3-\sqrt{10},0),(-3+\sqrt{10},0)\,;(0,-1)$

13. $(-2,0),(3,0)\,;(0,6)$

Section 11.7

Objective a

1. \$6; March 2007

3. 21 ft by 21 ft

5. 5 ft by 5 ft

7. -25; 5 and -5

Objective b

9. $f(x)=mx+b$

11. Neither quadratic nor linear

13. $f(x)=\dfrac{5}{3}x^2+2x-\dfrac{2}{3}$

Section 11.8

Key Terms

1. polynomial

3. test points

Objective a

5. $\left(-1,\dfrac{9}{2}\right)$

400

7. $(-\infty,-1]\cup[4,\infty)$, or $\{x \mid x \leq -1 \; or \; x \geq 4\}$

9. $\varnothing$

Objective b

11. $(3,\infty)$, or $\{x \mid x > 3\}$

13. $[-3,-2)\cup[5,\infty)$, or $\{x \mid -3 \leq x < -2 \; or \; x \geq 5\}$

Chapter 12 EXPONENTIAL AND LOGARITHMIC FUNCTIONS

Section 12.1

Key Terms

1. exponential

3. decreasing

Objective a

5.
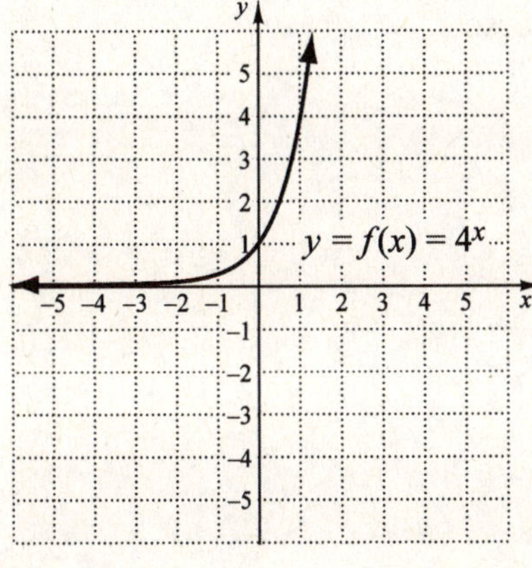

7.
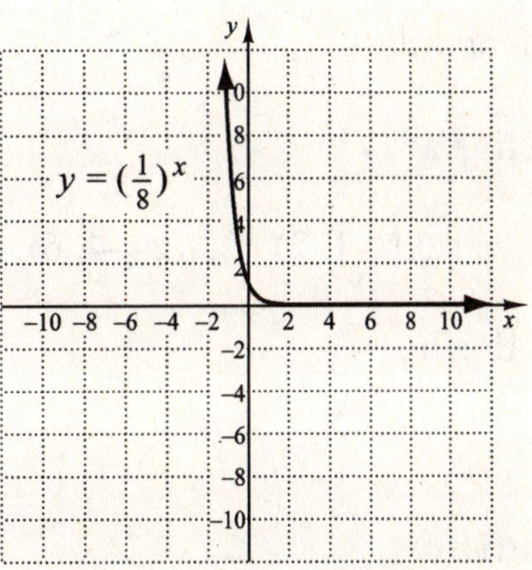

Objective b

9.
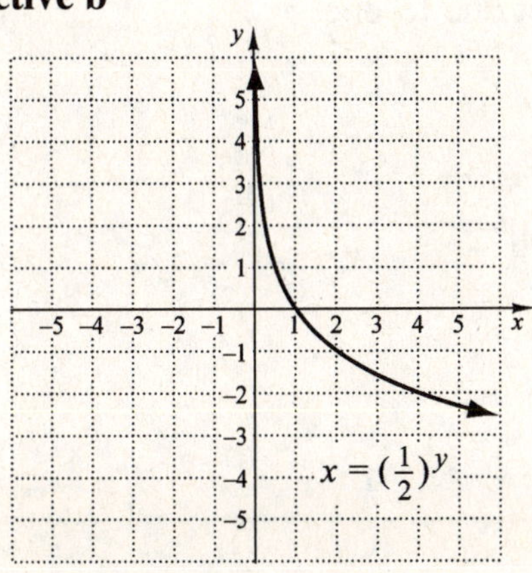

11.
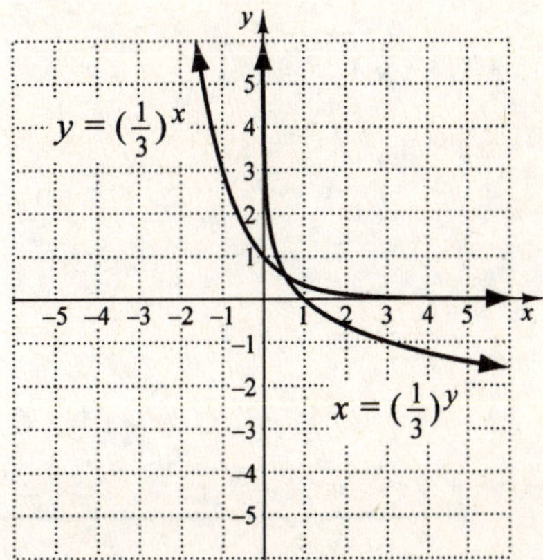

Objective c

13. a) About $12.4 billion;
     about $798 billion;

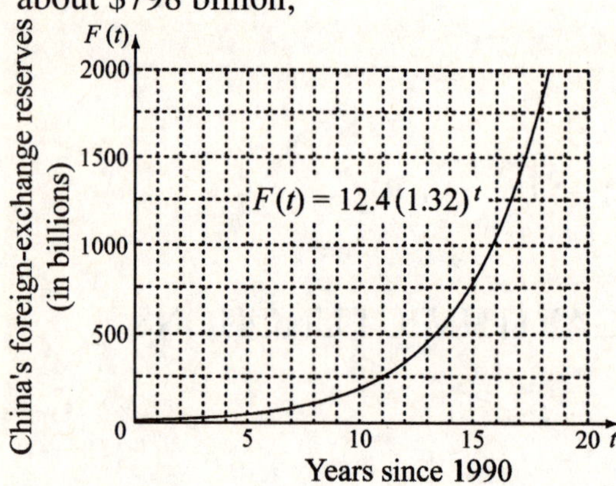

Section 12.2

Key Terms

1. composite

3. inverse

Objective a

5. Inverse: $\{(-1, 2), (3, 6), (-4, 6), (0, 8)\}$

Objective b

7. Yes

9. No

Objective c

11. $f^{-1}(x) = x - 1$

13. Not one-to-one

15. $f^{-1}(x) = \sqrt[3]{x - 2}$

402

17.

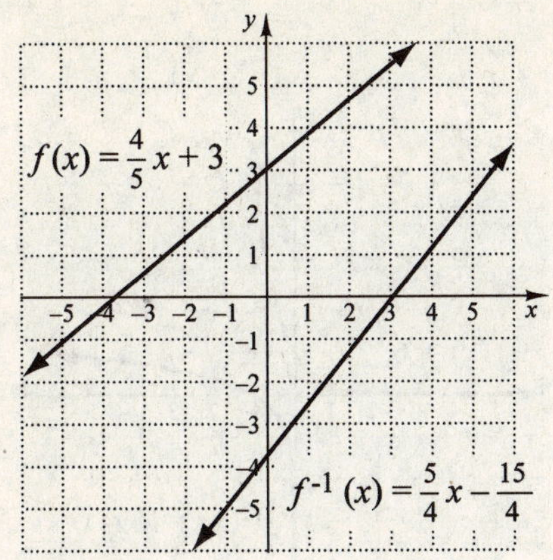

Objective d

19. $-1.38x - 8$; $-1.38x + 8.22$

21. $\dfrac{80}{x^2} + 2$; $\dfrac{4}{5x^2 + 2}$

23. $f(x) = x^2$; $g(x) = 10 - 3x$

25. $f(x) = \dfrac{4}{x}$; $g(x) = \sqrt{2x + 7}$

Objective e

27. (1) $(f^{-1} \circ f)(x) = f^{-1}(f(x))$

$$= f^{-1}\left(\sqrt[3]{x+2}\right) = \left(\sqrt[3]{x+2}\right)^3 - 2$$

$$= x + 2 - 2 = x;$$

(2) $(f \circ f^{-1})(x) = f(f^{-1}(x))$

$$= f\left(x^3 - 2\right) = \sqrt[3]{x^3 - 2 + 2}$$

$$= \sqrt[3]{x^3} = x$$

29. $f^{-1}(x) = \dfrac{1}{4}x$

31. $f^{-1}(x) = \dfrac{1}{x}$

33. a) 5; b) yes; $f^{-1}(x) = x + 32$; c) 40

Section 12.3

Key Terms

1. logarithmic

3. common

Objective a

5.

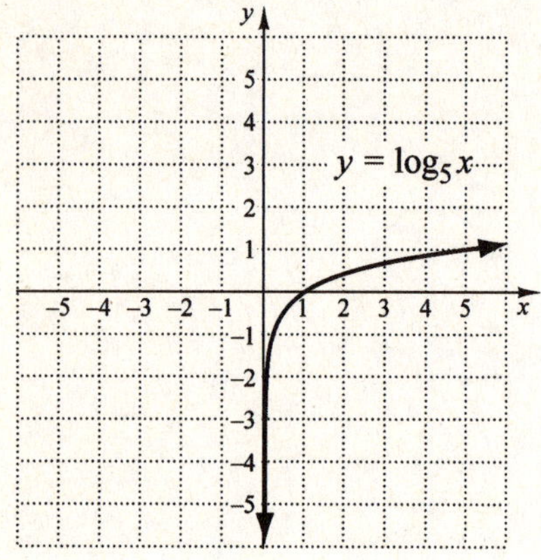

7.

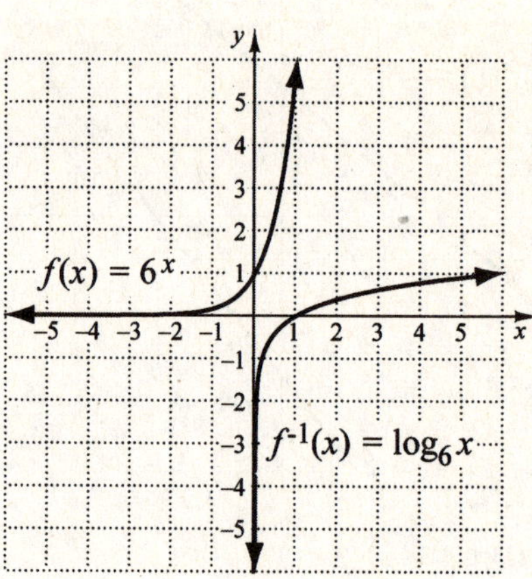

Objective b

9. $\dfrac{2}{3} = \log_{27} 9$

11. $-3 = \log_e 0.0498$

13. $a^p = t$

Objective c

15. 16

17. 1000

19. 5

21. -1

23. 0

Objective d

25. -0.5376

27. -0.3826

Section 12.4

Key Terms

1. c

3. e

5. f

Objective a

7. $\log_2 4 + \log_2 128$

9. $\log_a (9 \cdot 10)$, or $\log_a 90$

Objective b

11. $6 \log_a t$

Objective c

13. $\log_3 10 - \log_3 13$

15. $\log_a \dfrac{25}{11}$

Objective d

17. $\log_b w + \log_b x + \log_b y + \log_b z$

19. $3\log_a w + \log_a x - \log_a y - 2\log_a z$

21. $\log_x p^5 q^4$

23. $\log_a \dfrac{3y^6}{z^5}$

25. 1.183

27. -0.356

Objective e

29. 19

31. 6

Section 12.5

Key Terms

1. natural

3. domain

Objective a

5. 2.0794

7. 5.1160

9. 3

11. -5.2591

13. 1.7986

Objective b

15. 2.8614

17. 3.0276

Objective c

19. Domain: $\mathbb{R}$; range: $(0,\infty)$

21. Domain: $(2,\infty)$; range: $\mathbb{R}$

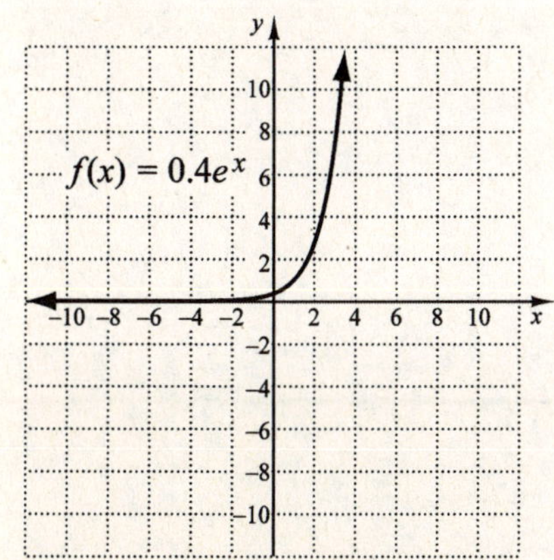

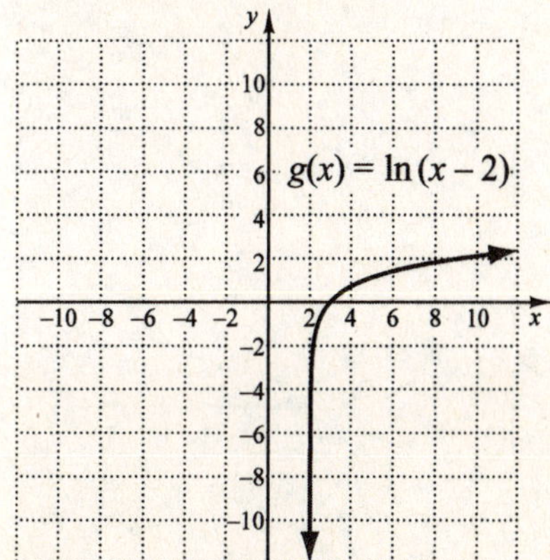

Section 12.6

Key Terms

1. exponential

3. exponential

Objective a

5. 3.4594

9. 0.3054

13. 48.1589

7. 4

11. 2

Objective b

15. $\dfrac{1}{9}$

17. $e^9 \approx 8103.084$

19. $10^{3/2} \approx 31.623$

21. 2

23. $\dfrac{251}{62}$

Section 12.7

Key Terms

1. decibel
5. decay

3. growth

Objective a

7. 98 dB

11. a) 72%; b) 63%; c) 55%

9. 10.5

Objective b

13. a) 2.7 yr; b) 17.7 yr

15. a) $P(t) = 28e^{0.03t}$, where t is the number of years after 2006; b) 31.6 million; c) 2018

17. About 1354 yr